ENCOUNTERS
WITH
WHALES & DOLPHINS

ENCOUNTERS
WITH
WHALES & DOLPHINS

WADE DOAK

SHERIDAN HOUSE

D# 36430

Library of Congress Cataloging-in-Publication Data

Doak, Wade.
    Encounters with whales and dolphins / Wade Doak.
        p.   cm.
    ISBN 0-911378-86-3 : $29.95
        1. Cetacea. I. Title. II. Title: Encounters with whales and dolphins.
QL737.C4D63 1989
599.5–dc19   28 00

Printed in Hong Kong.

To my wife Jan, who has typed, and collaborated on all ten of my books, one of the few who can read my hand-writing, and the best diving buddy, sailing skipper and partner one could wish for.

'I don't understand,' said the scientist, *'why you lemmings all rush down to the sea and drown yourselves.'*
'How curious,' said the lemming. *'The one thing I don't understand is why you human beings don't.'*
James Thurber

Acknowledgements

If there is a magic on this planet, it is contained in water.
Loren Eiseley

To acknowledge the immense assistance I have had in the ten years preparing for this book would occupy a great many pages. Now that Project Interlock operates four international networks the best I can do is thank people country by country: from Argentina to Australia; Brazil, Canada, Eire, France, Germany, Netherlands, New Zealand, South Africa, Sweden, United Kingdom and the USA, we have received unstinting generosity and heart-warming personal trust. For us all, dolphins and whales have made the planet seem smaller, more vulnerable and more precious to safeguard.

Contents

Prelude

Nature will deliberately reveal itself . . . if only we look.
Thomas Alva Edison

Since our study of dolphins in the wild began my wife Jan and I have had the most enthralling adventures of our diving lives.

In May 1984, I concluded my tenth book, *Ocean Planet*, with the words: 'Larry Vertefay writes to tell us he has just purchased a 60-foot catamaran for dolphin research; would we like to join his group for six months in the field, in a major effort to span the gulf between the oceanic and the terrestrial mind.' With that, my diving autobiography went to press and Jan and I set out for Florida. As we flew away from our New Zealand home we had high hopes, but what actually happened to us in the Bahamas aboard RV *Dolphin* exceeded every one of them.

On the edge of the Gulf Stream, within the Bermuda Triangle, we met a certain group of Atlantic spotted dolphins, the most approachable in the world, in company with a superb range of aquatic Americans of all ages. Our understanding increased by a quantum leap.

But those events, and what followed back in New Zealand, can only be related in the context of how these interspecies adventures all began, as much for Jan and me as for the hundreds of people around the globe who have been encountering dolphins and whales on their own terms. We call it *interlock*.

The initial five years of our dolphin encounters were recorded in detail in *Dolphin Dolphin*. Certain aspects of these encounters are relevant to patterns of understanding that emerge in this sequel. In essence, our early initiatives met with encouraging responses in several modes of communication. The dolphins showed they would prolong contact with us if we acted communicatively towards them; a valuable discovery if they were ever to be understood.

Dolphins also indicated they would extend their signalling systems to us, if we first established a receptive context. From exchanges of simple acoustic signals, we progressed to more subtle channels: mimicry extended to gamesplay and certain gestural signals emerged such as the 'bubble-gulp' which seem to have significance to dolphins of several species and in different parts of the world.

Our first major experiment, the dolphin suit, was an elaborate attempt to mimic their body language. In years to come, I was surprised to find such an odd approach had been suggested by an eminent scientist, Dr Donald Griffin.

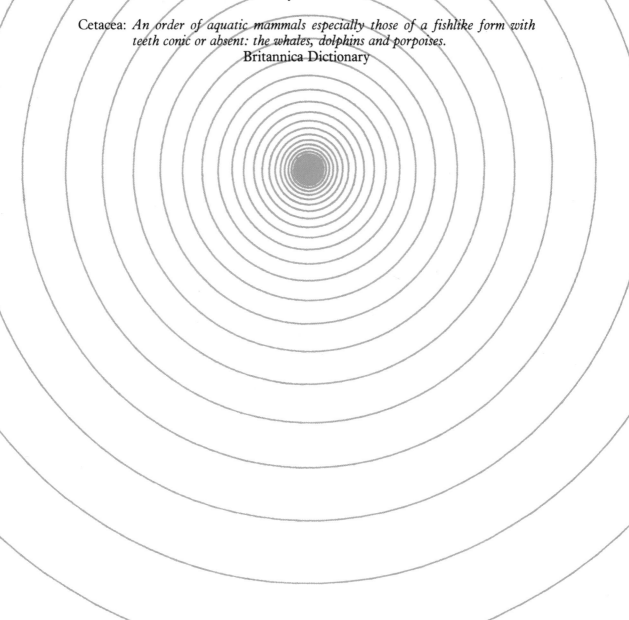

PROLOGUE

Quest for Ocean Mind

In terms of cortical size, degree of folding and cellular organisation, dolphins are the most highly evolved in the whole animal kingdom, and are, in every anatomical way, comparable to ourselves.
Dr Lyall Watson

Cetacea: *An order of aquatic mammals especially those of a fishlike form with teeth conic or absent: the whales, dolphins and porpoises.*
Britannica Dictionary

In the following pages I shall give an account of my first, most unexpected meeting with dolphins in the open sea. That episode set Jan and me on a quest which now spans a decade and two oceans.

After 10 years spent in research, I wonder how much further I will go before answering the questions that first encounter aroused. What are the capacities of large-brained aquatic mammals? What is the quality of their mental experiences? Could convergent evolution have produced an awareness, in the sea, with some potential for relating to large-brained bipeds such as ourselves? Could communication provide us with a window into ocean mind?

Meanwhile, I have attended two international conferences which considered these questions.[1] I have met curious humpback whales off Cape Cod, amiable scientists in Boston and Washington DC and four kinds of dolphins in the Bahamas, Hawaii and New Zealand. I have corresponded with thousands of people who have met dolphins and whales. I have had global feedback from my first book on this subject[2] and the 46 newletters we have published per our 'Project Interlock'.

From all this, I now reflect: if a person could become detached from a species-centred viewpoint and move to some hypothetical 'other' position, then review the documentation of human/cetacean approaches that have taken place in open-ended situations, perhaps some of my questions would be answered with quite profound implications.

The greatest difficulty is to avoid distorting the picture through the lens of human perception with its potential for unbridled anthropomorphism or giving human characteristics to other creatures, on the one hand, and on the other, the Cartesian extreme whereby animals are seen as reflex machines with no mental experiences whatsoever — or none that could be accessible to human investigation.

Coming to grips with the possibility of a non-human mind involves a delicate balance between the credulity of those who readily deify dolphins (as has happened in the past) and the institutionalised scepticism of some scientists who identify with the mindset of a generation that subjected cetaceans to captivity, invasive experimentation and weapons research, while larger species were being hunted close to extinction.

In considering the potential of the cetacean mind, I have encountered stumbling blocks. There is a curious linkage between science and religion: both a high church cleric and a field biologist responded with equal scorn to our study. To both, it represented a fundamental challenge to the existing order — a *heresy*!

Compared with most terrestrial mammals the field study of fast-moving ocean mammals is a 'Mount Everest' for science. Ocean research is more expensive and difficult to accomplish than that of a shore-based laboratory. There are really only two pathways for learning about cetacean living patterns — either through passive observation, such as the cliff-top studies of dolphins and whales,[3] or participatory investigation whereby communication may be used as a window into the alien mind.

The likelihood of reputable scientists receiving research grants for study of cetaceans would be greatly enhanced if the project were oriented towards captive subjects and classified military projects. As a result, human understanding of our closest brain rivals has been marred by secrecy and pragmatism. Military scientists are unlikely to give cetacean subjects the mutual respect and sympathy needed to develop an adequate two-way communication model.

Ironically, just as our tool-using species faces the prospect of a self-inflicted nuclear winter, we are on the verge of proving (Dr Kenneth Norris has stated he is 80% certain[4]), both from the fossil record of jaw development and observed feeding patterns, that some species of toothed cetacean (dolphins and whales) may have evolved a lethal biological weapon. It is possessed by each individual but never used other than for stunning prey; not even against their greatest predator — man. Understanding such a powerful ethical code may be the most valuable lesson we could derive from cetacean studies.

Despite numerous barriers, scientific evaluation of the cetacean brain has now made considerable progress, and the most qualified scientists in this field have little doubt that an exceptionally complex biological computer parallel to those of terrestrial mammals *but millions of years earlier*, has evolved in the ocean.

Because certain experiments[5] of questionable ethics were carried out on living dolphins, scientists actually know more about dolphin brain anatomy than they do about our own. A review of current neuro-anatomical opinion in this field provides a firm baseline for investigation of cetacean responses in the wild. New evidence supports the view that a big brain *does* correlate with advanced neuro-behavioural qualities. Dr Myron Jacobs, with colleagues Dr Peter Morgane and William McFarland, has made a comprehensive atlas of the dolphin brain. At the 1980 IWC Conference on Cetacean Intelligence held in Washington DC, I heard Dr Jacobs, and others, present their views on the architecture of the cetacean brain.

Dr Jacobs said: 'The great expanse of association cortex strongly suggests that cetaceans have a highly developed but different form of intelligence.'[6]

Dr James Mead, Curator of Marine Mammals at the Smithsonian Institute, provided a most pertinent point — rather than clouding the issue with arguments of equivalence or superiority, we should be 'seeking to understand what the cetaceans *do* with their large brains'.[7]

The cetacean brain has been examined by the neuro-anatomists and has been found to be superb. But what do academics know of its performance?

We could begin by examining the record of behavioural research into captive dolphins, such as the celebrated experiment in Hawaii, where Karen Pryor[8] rewarded a dolphin for providing new behaviours and then documented its performance of so many new behaviours that the experimenters were unable to categorise them adequately.

Dr Sam Ridgeway[9] recorded the ERPS (Event Related Electrical Potential) from a dolphin brain and compared them with similar experiments in humans and monkeys. Only dolphins and humans were comparable in sharing properties of ERPS known to be 'decision' related. In short, the enlarged areas of the cetacean brain operate at levels of complexity previously found only in our own brains.

A thorough review of captive cetacean research would be inappropriate here. Such evaluations do corroborate the high expectations of the anatomists but still leave us on shaky ground. If we wish to comprehend truly the capacities of a large-brained social creature, it should be in a context where performance is as unbiased as possible. Only in a state of freedom can a creature manifest its full range of behavioural flexibility.

Scientific observers are doing their utmost to learn from the passive studies of field behaviourists, such as Dr Bernd Würsig. In relatively few areas (Argentina, South Africa, Hawaii) has it been possible to observe the social lives of coastal dolphins from cliff-top vantage points.

In the course of a four-year study of bottlenose dolphins in Argentina, Dr Würsig and his wife, Melany, gradually learnt to recognise most of the individuals in various subgroups. He then found there was a degree of openness in the groups — some individuals changed their association after a few days while five or six individuals swam together consistently for at least 18 months. When groups met after separation, they vocalised more than they ever had previously, usually stopping to exchange greetings, nuzzling and caressing.

Observing dusky dolphins feed cooperatively, he noticed that they took turns to go through the fish school to feed while others kept the school tightly packed. Würsig realised that such cooperation required highly refined communication, otherwise certain individuals might grab more fish and spend less time herding. He concluded that it was likely the dolphins knew and trusted each other enough to control the situation. As well as remembering each other, they appeared to have an extended concept of 'groupness' rather like our concept of belonging to a club or society which excludes constant physical association. Defining 'awareness' as being cognizant of one's actions, or thinking about past, present and future, Würsig asserts that we should admit the possibility of awareness in non-human mammals.[10]

Other field studies corroborate and extend Würsig's cautious suggestions. In the Hawaiian bay where Captain Cook was killed, Dr Kenneth Norris,[11] Professor of Natural History at the University of California, Santa Cruz, has set up a 14-year-old study of spinner dolphins. The steep-walled cove offers a rare opportunity to study dolphins in the wild because of the cliff-tops, the clear, calm water and the habits of the spinner dolphins themselves. This tropical species has a nocturnal feeding pattern and social groups spend the daylight hours resting and romping in large sheltered bays such as Kealakekua. Prolonged observation has now yielded some exciting discoveries, far beyond what could be learned from placing such social creatures in captivity.

To date, passive field observation has shown that dolphins have strong, extended mother-infant ties; that females have bonds with infants other than their own offspring, and that maturation is a slow process involving much learning and play behaviour. Their mating system involves rotating consorts without permanent pair bonds; dominant males tend to coordinate group activities, and there is evidence of coordinated group responses to danger and care of the dead.

Dr Norris asserts that these dolphins know each other as individuals, each emitting a separate whistle call; that they live in a complex, learned society where a dolphin may know as many as a hundred other dolphins; that they spend about one-third of their day reaffirming relationships through caresses and responses. He claims that their safety at sea depends on some kind of whistle network in which they maintain contact with all members of the school, modulating the whistle if danger threatens.

Norris concludes that dolphins are a high order of animals with a more complex social structure than a simple set of family ties; one that functions more like our own society where we have friendships and associations beyond the family. He points us to the prophetic words of Dr Gregory Bateson (1966), eminent anthropologist and ethologist:[12] 'My first expectation in studying dolphin communication is that it will prove to have the general mammalian characteristic of being primarily about relationship. This premise is in itself, perhaps, sufficient to account for the sporadic development of large brains among mammals.'

One day a whale researcher who had visited our project in New Zealand

and understood the direction of our study, sent me a small book titled *The Question of Animal Awareness* by Donald R. Griffin.[13] An experimental biologist at Rockefeller University, New York, Dr Griffin is best known for his discovery that bats and other animals use echolocation to orient themselves and locate food.

Few books have given me more inspiration. During ten years of exploring this field I had often felt isolated and vulnerable. In approaching any discussion of intelligence, mind and awareness, I knew I was entering a minefield, a highly controversial frontier of knowledge. Subtitling his book 'Evolutionary Continuity of Mental Experience', Griffin makes some definitions which I find useful in calibrating my thoughts about mind and consciousness in the ocean. For Griffin *mental experiences* are thoughts about events and objects remote in time and space from our immediate sensory inputs; *mind* is something that has such experience; *awareness is* the whole set of interrelated mental images of the flow of events, as immediate in time and space as the toothache or as remote as the expanding universe concept. *Consciousness* is the presence of mental images and their use by an animal to regulate its behaviour.

Griffin then goes on to review the evidence in areas of animal communication from bees to chimpanzees. He concludes that their use of versatile communication systems is evidence of mental images and a capacity to communicate with conscious intent.

By that stage (1984), with the evidence from reputable scientists of dolphin awareness, and Griffin's definitions, I felt much happier about my own thesis that the ocean creatures I had been meeting possess awareness of a quality yet to be established; that ocean mind exists and may communicate with us if we can establish appropriate channels. At the same time I was aware of the danger that this belief could be exaggerated, to the horror of scientists who dread the popular image of dolphins as 'humans with flippers'.

We must not jump the gun, however. In accepting that dolphins and whales probably have mental experiences of a high order, we must not assume that they are identical with our own. We know so little. We are just beginning to perceive the first hints of ocean mind. The implications of what we know already are so profound that it may take a century for it to sink in and even longer before our species can really come to terms with it. In the meantime we should heed Dr Lyall Watson's advice: 'Allow for the animal's awareness, but do not make the mistake of assuming it will be similar, or even comparable, to your own.'[14]

In his final and most controversial chapter, 'A Possible Window on the Minds of Animals', Griffin outlines suggestions for the exploration of a scientific territory so unknown that its existence has been seriously questioned. The anthropomorphism taboo has long made it dangerous for any ethologist to consider that animals have mental experiences. This dates back to Descartes who regarded animals as mere machines. To this day, some scientists consider them as 'prewired' genetic programmes.

Griffin suggests that, since animals *do communicate* with each other, perhaps we could learn something of their minds if we approached them in the way that an anthropologist[15] studies a group of people whose language he does not know. With communicating animals, the investigator might talk back and forth, perhaps *through an appropriate model*[16] to verify the meaning of its communication signals. Griffin then outlines a novel approach for establishing a two-way exchange which he terms 'participatory research into interspecies communication'.[17]

When I read these words after years in the wilderness of intuitive gropings towards such goals, I felt I had come in from the cold. Here was a formal exposition of the things I had been attempting: a style of approach for channelling appropriate communicative gestures.

To explore subtle aspects of the body language of the honey bee, he suggests using a model bee to interact with real bees; or impersonating a chimpanzee with a thorough disguise including appropriate sounds and pheromonal perfumes. (Our dolphin suit experiments parallel these proposals although the technology involved in modelling dolphins accurately is much more demanding and expensive. We had no intention of deceiving the dolphins.)

Such efforts would meet with ridicule in some scientific circles but Griffin presents a long roster of researchers who, for the past 50 years, had attempted to deal with the complexities of the animal mind in a disciplined manner, trying to steer a course between the extremes of anthropomorphism and the Cartesian reflex machine. Of the new genre of researchers he states: 'First they must overcome the feeling of embarrassed outrage at this notion and then laboriously develop the necessary techniques of disguise, imitation and communicatory interaction.'[18]

He foresees that the researcher might experimentally control messages until he understands their effective content. He may be able to ask questions and receive answers about an animal's mental experiences. A major objection then arises — the mental experiences of other animals may be so different from our own thoughts that we cannot recognise them.

Griffin takes comfort from the evidence of physiologists that the nervous system and neurons of all multicellular animals are basically similar and that an evolutionary kinship exists between animals and humans — 'Neurophysiologists have so far discovered no fundamental difference between the structure or functions of neurons in men and other animals.'[19] Anthropomorphism he calls an obsolete straitjacket.

After I read Griffin's book, my quest for a context into which an understanding of ocean mind might grow met with another stroke of luck. At the 1980 Conference on Cetacean Intelligence in Washington DC, I met psychologist Dr Michael Bossley of Magill University, South Australia. Later he sent me an extraordinary unpublished manuscript — his review of the scientific evidence for non-human mind, which was a global survey of formal research into cognitive ethology since Griffin had defined it. I read this with utter delight and suggested a title, *Continuum*, which Dr Bossley accepted.

The implications of Bossley's survey could upset many. He insists that an entirely new ethical system is required, and presents compelling evidence for a continuity between human psychological processes and those of other life forms. He urges our species to climb down from its imaginary pedestal: 'Everything grades into everything else. We *are* part of the natural world.'

Much of the research Bossley examines is recent and ongoing. For the most part it has appeared only in highly technical literature accessible to specialised academics. It may be several generations before the full implications are heeded. Like the Copernican and Darwinian revolutions, it could alter the way we view our place on this planet, how we treat other life forms and each other.

Legitimate evidence that five vital aspects of being human can be traced to other animals exists in the published work of established scientists. In each of five chapters, Bossley summarises that evidence. The presentation I am about to make of our own research into human/cetacean relationships belongs in this context — the continuum of mind that extends into the ocean and forests of this planet. I do not wish to place cetaceans on a lonely pedestal adjacent to our own but rather, to provide hard-won evidence from the sea that extends and reinforces both Griffin's and Bossley's theses. I suggest that we visualise the mind continuum not as a hierarchy or ladder ('The Great Chain of Being'), but as degrees in a compass rose.

For the great scholar, Gregory Bateson, 'mind' is a network of interactions relating the individual to his society and his species; 'ideas' develop and evolve according to the same laws that control natural phenomena.

In his collected essays, *Steps to An Ecology of Mind*, Bateson wrote: 'The individual mind is immanent but not only in the body. It is immanent also in pathways and messages outside the body; and there is a larger mind of which the individual mind is only a sub-system.'[20]

In this book I propose to review ten years of my personal experiences with dolphins and whales, along with those of other people, in close approaches ranging from mutual curiosity to gamesplay, mimicry and complex interaction. *Dolphin Dolphin* included my study of altruistic encounters where dolphins have protected people, rescued them or warned them of danger. Although space precludes further coverage of this aspect in this book, our files have expanded considerably since 1981.

Long regarded by scientists as rigid, stereotyped or capable of innate responses only, the care-giving or *epimelitic* behaviour of dolphins has recently been reappraised.[21] The emergent picture reveals a long-lived, complex and mutually dependant dolphin society, involving extended parental care, cooperative feeding, and an extremely fluid social structure. Within such a society — closely paralleling our own — behaviour is typified more by learned than innate responses. When swimming with unrelated 'friends', mutually assured assistance is clearly important. As with humans, selective pressure for more and more sophisticated acts of altruism can be expected. Furthermore, there is much evidence that dolphins and whales extend their care-giving and cooperative behaviour to species other than their own. In this context their assistance to humans becomes all the more credible.

Related to the altruism of dolphins towards humans are the well-documented (filmed) episodes of dolphins assisting fishermen. Such commensal fishing is not necessarily altruistic, however. In many cases both species appear to benefit, but it can be seen as part of a continuum of cooperative interspecific behaviour and offers further insights into the nature of the interspecies bond, and the surprising capacities of cetaceans to interact with us in open-ended situations.

Anecdotal Evidence — For too long accounts of friendly or altruistic cetacean encounters have been dismissed as folklore. Because such episodes may occur only once in a lifetime and credibility is at risk, many people become reticent. This further isolates those who chose to speak out. Even though these incidents are rare, they are hard-grain reality. I have made a major effort to document and collate them, and claim that such anecdotal material should be considered in any evaluation of the cetacean brain.

Observation of open ocean behaviour is exceedingly difficult for scientists but the accounts I have amassed do not bear this out in all respects. Where people have behaved towards cetaceans in a benign, communicative manner they have often met with prolonged and remarkable responses. Yet, we must remember that brain anatomy, captive and field studies, all concur in respect to the quality and potential of this ocean mind. Because close approach situations appear anomalous and have no relationship to scientific evaluation, they are often dismissed as irrelevant. But this is not always correct. Accounts given by lay people fit neatly into the scientific assessment of the cetacean brain. The behaviours described may be uncommon but they are what might be expected of a large-brained social animal in a communicative setting with its closest brain rival. Investigation of advanced, non-human minds is a novel field for western science. After centuries of species-centred bias, it is going to demand unconventional adaptation of scientific methods. The Russian scientist A.V. Yablokov believes it may be impossible for us to understand an alien, non-human thinking system from current anthropocentric research methods based on the premise that man is the centre of the universe.[22]

Lone Dolphin Encounters — During the course of our studies we became aware of a special category of human/cetacean encounter — the situation where a lone dolphin spends an extended period of time around human settlements. In many cases its normal social intercourse seems to have been replaced by intensive interaction with people. Such episodes appear to have increased in recent years, perhaps facilitated by the change in attitude towards dolphins — an account from last century culminated in the dolphin's capture and display on a hand cart! For want of a better name I have labelled these encounters *Dints* and now have a file of lone (though there have been pairs and even sub-groups) dolphin/human relationships.

Obviously these episodes offer little knowledge of the dolphin's normal social life but they do complement, in some respects, dolphin school observations, and they provide unique insights into the flexibility and complexity of their relationship with an alien species which does not share their acoustic, non-manipulative culture. By amassing a range of such accounts from all over the globe, certain patterns emerge. Simplistic explanations of the phenomenon, based on too few examples, are shown to be unlikely.

Friendly Whales — Considering that a dolphin is really a small-toothed whale, it is not surprising that friendly encounters have increased among the more common of the 76 cetacean species. As the whale killing industry winds down, populations are showing a promising resurgence in many parts of the world — in their former feeding grounds and nursery areas. The situation has changed since the days before industrial man declared war on the cetaceans. For the first time in history many whale grounds have been given sanctuary status. We now have the technology which enables people to meet whales on their own terms and to listen to whale voices. The understandable fear of close proximity to creatures many tons in weight, is proving groundless as we learn the appropriate ethics for meeting leviathan in a benign setting. And so our files have accumulated accounts and photographs of close approaches involving humpback, grey and sei whales; minke, Brydes, fin and right whales; sperm and pilot whales; orca, pseudorca and beluga — all offering further insights into the capacities of ocean mind.

NOTES

1. 'Cetacean Intelligence & the Ethics of Killing Them', Washington DC 1981. 'Whales Alive', Boston 1983. Both held under the auspices of the International Whaling Commission.
2. *Dolphin Dolphin*, Hodder & Stoughton (1981 Auckland) and Sheridan House (1982 New York).
3. Drs B. Würsig, R. Payne et al.
4. 'Do odontocetes debilitate their prey acoustically?' Abst. Fourth bien. Conf. Mar. Mam. San Fran., 1981.
5. Drs P. Morgane et al. 'The Whale Brain & the Anatomical Basis of Intelligence' (Scribner, NY 1974).
6. M. Jacobs, 'Studies on Cetacean Brain'. (Paper at Conference, 1981.)
7. J. Mead, 'Whalewatcher'. (Fall 1985, No 3.)
8. K. Pryor et al. 'The creative porpoise: training for novel behaviour'. (Jour. Anal. Behav. 12: 653-661.)
9. S. Ridgeway, 'Cetus' 3/5, p4.
10. B. Würsig, 'The Question of Dolphin Awareness Approached through Studies in Nature'. "Cetus" 5-1, pp 4-7.
11. K. Norris et al. 'The Behaviour of Hawaiian spinner dolphins *Stenella longirostris*'. Fish. Bull. 1986, 77(4), 821-49.
12. G. Bateson, *Steps to an Ecology of Mind*. (Chandler Pub Co. 1972 p337.)
13. D. Griffin. 'The Question of Animal Awareness', Rockefeller Univ 1976. : 'Animal Thinking', Harvard UP (Cambridge 1984).
14. L. Watson. *Whales of the World*. (Hutchinson, London, 1981, p49).
15. Griffin, op. cit.p88.
16. Ibid. p95.
17. Ibid. p95.
18. Ibid. p95.
19. Ibid. p104.
20. G. Bateson, op.cit. p436.
21. 'Are Dolphins Reciprocal Altruists?' R.Connor & K.Norris. Am. Naturist, Mar. 1982, Vol.119/3, pp358-374.
22. A. Yablokov, 'Behavioural Difference between Species and Groups of Species'. (Comment at Conference, 1981).

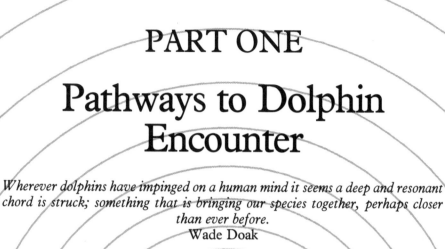

PART ONE

Pathways to Dolphin Encounter

Wherever dolphins have impinged on a human mind it seems a deep and resonant chord is struck; something that is bringing our species together, perhaps closer than ever before.
Wade Doak

Project Interlock Begins

Dolphins entered my life at a most appropriate time because a major phase of diving exploration had just been completed and I had photo files bulging with pictures of the myriad life forms in warm southern seas — but never a dolphin.

My photo collection had begun in 1959, when a hoard of silver coins salvaged from the treasureship *Elingamite* gave me the nest egg that led to full-time study of marine life. Until then I'd been a teacher with a language degree and a passion for the undersea world. Quitting the classroom, I devoted a dozen years to illustrating and researching books about reef fish behaviour and the ecology of invertebrate creatures on the undersea cliff. Then I wrote a further two books on the explorations of *El Torito*, the undersea research vessel belonging to American scientist, Dr Walter Starck. With him, and later my whole family, I spent several sundrenched years exploring coral reefs and their inhabitants, from polyp to atoll dweller, sharks and shark worshippers.

In fact, when I think back on it, the yarn I'd had with Walt in the saloon of *El Torito* probably triggered something which set me off in the direction of this book. We were riding out a hurricane in the lagoon at Lord Howe Island when this leading diving biologist began relating his life story for me.

Walt started with the formative childhood experiences in Florida Keys where his father was a charterboat skipper. 'This day my father caught a baby dolphin for a dolphin pool. The mother stood by our boat making such piteous sounds that my own mother called on him to release it. This must be the only animal that has the power to kill a man, but *always* forbears to use it, even in a situation like that. Among dolphins themselves, there just doesn't seem to be any aggression and they accept man on the same level.'

For some reason I always remembered Walt's avowal that, if he ever had the means, he would devote himself to solving the dolphin enigma. I was, and still am, inspired by that ambition.

In all my diving life dolphins had only been peripheral to the undersea world I was exploring. For most divers around at that time it was, typically, a hello-goodbye type of encounter. Dolphins fled if we tried to jump in with them from a boat. I'd only once had a longer acquaintance.

In May 1971, at the Poor Knights Islands, 22 kilometres off the east coast of northern New Zealand, I had been 20 metres down where white sand lies in drifts at the foot of the Rikoriko cave wall. Suddenly everything went black. I glanced up in fear. The cave portal, usually a blaze of blue fire, had dimmed. It was seething with huge shapes — sharp fins, fast tails and jaws. Dolphins. One of them, in silhouette on the surface, smacked its tail and a pair of curving forms glided down in a spiral, circling me and rising. Each time the leader of this game smacked with his tail a pair spiralled down, curving on their sides to gaze intently as they passed me. Enchanted, I forgot for some time that I had a camera. Then with the last two frames left on the film I took my first dolphin pictures — *Tursiops truncatus*, the bottlenose. A seed was sown.

After our *El Torito* adventures, my family settled down to a rural life in a small cottage beside a mangrove river in Northland, New Zealand. It was one evening in April 1975, after a superb day's diving at the Poor Knights Islands, that I told Jan, Brady and Karla of a most unusual experience (little knowing how much it was to change our lives).

Along with biologists, Barry Russell, Tony Ayling and two other companions, Cathy Drew and Les Grey, we were returning to the mainland in my runabout when we met up with a vast school of bottlenose dolphins, four miles off Tutukaka Harbour. Our 'deep vee' was bucking over a switchback sea, running so fast we clipped the wave crests, virtually airborne. The five of us were totally exhilarated after the day's diving. Suddenly, there were dolphins ahead of the boat. They raced to meet it, arching from the water as they rose to breathe, making their presence known. I eased the machine to whale speed and the dolphins adjusted their tails to its pressure wave. They surfed ahead of us, cavorting from port to starboard, rolling on their sides to eye the glassy white hull and its occupants hanging over the bow shouting encouragement and slapping the hull with staccato rhythms. Both species, man and cetacean, had begun gamesplay.

Looking back, I feel that I acted in a premeditated way. I recall that our fast ride had begun in the cathedral-quiet green twilight of Rikoriko, 'Cave of Echoes', the 25-metre-high dome inside Aorangi, chief of the Poor Knights. We had taken Cathy, the New York photographer, in there to view its vastness and had grown silent as we studied the vaulting roof with delicate ferns clinging to it, marvelling at the dim light they tolerate.

Now, as the dolphins moved to meet our fast-running boat, I knew exactly what I wanted to do when I met dolphins on the bow. I wanted to slow to a crawl, put the boat in a wide circle and leap into the centre with my camera. I wanted dolphin photos, at this stage with the thought that they would satisfy a publisher's request.

Tony took the wheel and I plunged in. My strategy worked — dolphins were frolicking around the bow like circus ponies, with me as ringmaster. I fired off my photos, then became aware that roles were changing. The boat had stopped and I was now at the centre of a cyclone of dolphins. I called to Tony and the others to join me, but one at a time, and 'let's all do the dolphin-kick'.

On island voyages with Dr Walt Starck we often entered villages where there was a language barrier. To our delight the kids would open up communication through playful mimicry and dance. During our dolphin games each diver in his sealed-off world became aware that the dolphins were demonstrating new tricks. I was weaving among them with a fluid dolphin drive, my fins undulating together like a broad tail in a movement that began at my head and rippled along my body. A dolphin drew alongside me. By counter-opposing its flippers, like the ailerons of a plane making a spin, it barrel-rolled right in front of my mask. Maintaining the dolphin-kick, I imitated this corkscrew manoeuvre, counter-opposing my hands held close to my chest. The response was slow as my pseudo-flippers were tiny and my speed a fraction of theirs, but I found myself rolling wing over wing. Then something startling happened. The moment my spin was complete a formation of six dolphins abreast of me and on the same side as before, repeated that trick in unison, reinforcing my newly acquired mimicry pattern. And so it went on, the sea wild with energy, a maelstrom of dolphins, their shrill chorusing whistles dinning in our ears.

We gambolled with them for about an hour until utterly exhausted. Then one by one we hauled ourselves in over the stern. Around us we could now see salvoes of dolphins leaping singly or in symmetrical pairs for a mile in every direction. We must have met a whole tribe of bottlenose dolphins on their passage along the coast.

As we towelled ourselves warm I commented how marvellous it was that such huge, sharp-toothed animals, each as heavy and as fast as my boat, had not even buffeted us by their swirling movements. 'Only once,' said Tony Ayling who had been last on board. As he was approaching the stern, one dolphin rushed to within a metre of him and had stopped short, vertical in the water, with flippers flung wide as if imploring us to continue. We felt we had let them down. I gave the hull a resounding thump. From a short way off a tail smacked on the surface in answer. I thumped again. Eight times we exchanged signals, and that was it.

For the next year I devoured every written word I could find on dolphins and whales. Knowledge of dolphins, I found, was based almost entirely on captive animals. How much could be learned about our species from a study of people in prison? Jane Goodall had shown what could be done through behavioural fieldwork with wild dogs and chimpanzees and proved that much more could be learned about these creatures when they were studied in their natural surroundings. But gaining acceptance of terrestrial mammals is far easier than with ocean-dwellers.

I discovered that there had been many problems when first attempts were made to communicate with dolphins. Back in 1969 when he wrote *The Dolphin: Cousin to Man*, Robert Stenuit pointed out that wild dolphins refused to associate with man underwater. For the most part, when divers encountered dolphins swimming free in the ocean and attempted to film them, their plans were cut short. The dolphins swam away. In his book, *Dolphins*, Jacques Cousteau openly admits defeat in his attempts to study dolphins in the wild. In order to complete a film, Cousteau had to capture dolphins and restrain them in the sea with nets and buoys. In a storm two net-confined dolphins were almost drowned. Cousteau declared he would never again capture dolphins to film them.

Gradually, during this year of library research, my ideas formed a pattern of possibilities to be explored. Could Jan and I take up the challenge those dolphins had offered with their body language and gain some insights into the enigma surrounding them? We had seen a way through the problem that had faced Cousteau, but where would it lead? Was mimicry and play the first step as it had been when we were meeting island people? If we remained passive observers, dolphins (or people) would leave after mutual scrutiny. Any advance towards them led to an early departure. But by gaining their interest through interaction we could possibly prolong contact which would enable us to recognise individuals and begin to understand them. Any study of free-ranging dolphins would have to be on *their* terms. The challenge would be to discover just what those terms might be. If we were sensitive to their behaviour an interspecies ethical code might emerge.

The charge that it would be anthropomorphic to study dolphins through the use of a communication model seemed absurd to me in view of the approaches those dolphins had already made to us, and because of the problems faced by conventional methods of study. But it would be a strange and unsettling process — a step beyond the scientific model of reality . . . Project Interlock had begun.

Body Language

The corkscrew manoeuvre and the mimicry of that first dolphin encounter provoked me to take body language much further. I had already made a series of behavioural experiments using models on fishes. I'd found that there are special markings and signal patterns which reef fishes use* in courtship, aggressive display and parasite removal. Certain ritualised movements or dance routines communicate messages between members of the same species and from one species to another. A black angelfish turns its back and extinguishes the white ear marking on its head to appease a dominant invader of its territory. Many wrasses bear signal flags on their fins, and some have signature markings on their bodies for personal recognition. Coral shrimps wave their six long, white feelers from beneath ledges to attract customers for parasite removal. The sabre-tooth blenny mimics the swimming pattern and colouration of the cleaner wrasse to approach unsuspecting fishes and nip pieces of their skin.

Assuming that dolphins are at least as intelligent as fishes in observing such signals and body language, what would be their response if I were to mimic them in colour, form and movement — to turn myself into a model dolphin?

I reasoned that something beyond all our previous experiences must have stimulated the intensity of our meeting and served to prolong the dolphins' interest in us that day returning from the Poor Knights. There had to be some underlying factor which would explain this spontaneous happening.

I decided to take it all one stage further. I would try the same performance again, but this time the humans would have dorsal fins fitted to their weightbelts. They would wear special wetsuits modelled on the dolphin body; black above and white below, both legs enclosed in a sheath of neoprene and terminating in jetfins or, ideally, a rubber tailfin.

As swimming bipeds, we instinctively use an alternating kick, but the sinuous waving of body and tailfin is a much more effective and widespread mode of water propulsion. Instead of increasing the power of our kick with rubber fins, why not copy the fish, as the cetaceans have done? This would be an indication to the dolphins that we too are intelligent observers of convergent evolution and have not permanently overlooked the fish's advantages. Propelled by transverse waves flowing through its musculature, the fish penetrates the sea with minimal disturbance or cavitation. This creates a highly efficient transfer of body energy to wave motion and minimises drag, which the dolphins have emulated.

Midway along the dolphin's body, the dorsal fin serves as a keel, providing lateral resistance about which the body can turn or pivot. Its function can be seen when the bow riders suddenly swerve aside and leave the ship. Such a fin attached to the centre leadweight on my belt, might assist 'dolphin-swimming'. Certainly it would help me get the feel of it while watching them.

*Described in *Fishes of the New Zealand Region*, W. Doak, Hodder and Stoughton, Auckland, 1972.

My whole body would have become a swimming organ.

The dolphin's flippers serve as ailerons for rapid diving, ascent and spiral turns. They are also important for communicating through touch, an essential part of social activity. I would wear a pair of hand fins framed with wire and covered in foam neoprene. I made a rough mock-up of my dolphin suit — just enough to test its practicality for swimming.

On a calm spring day Jan and I boat-bounced 22 kilometres over the southeast swells towards Aorangi, Island of the Cloud Chief. Two puffs of vapour, the only ones in the entire sky, hovered above the Poor Knights. Recent rains had discoloured the sea as far out as five or six kilometres — but the ocean around the islands was sparkling clear, a fresh upwelling after the storm. There were no signs of bird activity — just huge rafts of petrels bobbing on the swells. The new water was low on plankton life as yet. We cruised around the cliff faces and anchored for lunch on Landing Bay Pinnacle, an ideal place to test the gear.

Getting into the water presented special problems without the use of hands and legs. I set myself up on the stern, steadied my mask with one flipper and plunged headlong in. The momentum of my fall carried me completely through the surface and I found myself flying around the boat in an effortless arc. Thinking back on it the sensation was curious — as though I had left my body. My mind was still, as though contemplating the body's action from apart, much like the first time I flew a hang-glider. The dolphin motion was autonomous; I didn't seem to be consciously striving. Some cellular blueprint for undulating movement was, perhaps, triggered by the 'memory' extensions I had made to my form. In a situation where survival would instinctively depend upon a kicking action I was a legless monopode. Because my legs were hobbled, I had to move like a dolphin all the time and this forced me to perform better. The one-piece dolphin suit was a makeshift outfit I had glued up from neoprene scraps. It really needed to be properly tailored to preserve body warmth.

The dorsal-fin keel attached to the centre leadweight of my belt fitted snugly to my body. This made it easy to move around in curving pathways, and stabilised yawing movements, with a consequent saving in energy. The snorkel enabled me to breathe like a dolphin. I was well ballasted and soon I could glide around beneath the surface using my new body to rise and descend, turn and twist like a fish.

But the stinging cold sank in and I found myself losing heat too rapidly to maintain the pleasurable aspect of the performance. The more I moved, the colder I became — a punishment that spoilt the previous, and essential, spontaneity of movement. Approaching the boat to climb aboard, I learned how hard it is to tread water with one leg and no arms.

I was convinced. As soon as it could be afforded, I would have a proper dolphin suit and appendages constructed. In the meantime, the dorsal fin would be a permanent fixture on my weightbelt. I liked the improvement it made to my ability to move like a dolphin. Next time we meet, I would mimic them better.

Almost a year had elapsed since that initial cavort with the dolphins and we had gone to considerable lengths to prepare ourselves for another encounter.

One morning in November 1976, Jan and I headed the Haines Hunter out of Tutukaka and wave-danced towards the Poor Knights. This time in readiness, I was wearing my snug-fitting, Farmer John wetsuit pants and all my gear was close at hand.

Our bow became flanked with curving shapes — common dolphins, *Delphinus delphis*. Jan took the wheel while I got the cassette-player going and cetacean sounds, *Tursiops truncatus*, played through a speaker in our bow. The boat circled slowly. I was scrambling into my gear, the dorsal fin on my weightbelt all set to don, when the dolphin sounds ended so we switched to the song of the humpback whale.

While I was in the water Jan was steering the boat in circles. I could see dolphins weaving everywhere, around me and below — silver black bullets in the green haze. Visibility extended to about seven metres because of the plankton bloom but groups kept whizzing by within ten feet of me, twisting on their sides, or leaping out of the water and plunging back. From time to time the numbers around me would disperse, and eventually, after about 15 minutes, contact was lost. Jan pointed to the bird activity a few hundred metres away — the dolphins had resumed feeding.

I climbed over the stern and we moved nearer to the feeding frenzy. Again the dolphins responded to our presence with frolicsome behaviour, leaps and tail-slapping. I slipped back in, began diving down, circling about and was surprised at the frequency with which they were defaecating right in front of me. Was it a signal?

As a dolphin passed me I heard an extra loud whistle and saw a string of bubbles emerge from its blowhole. I took the snorkel from my mouth and screamed 'ruuaark' at the bunch as they headed directly at me. They veered slightly and passed at close quarters, turning on their sides to eye me as they zipped by.

This session was much longer and more intensive than the first. It was full interlock in that no other stimulus than my presence was now keeping them from their feast.

After about 20 minutes the numbers around me diminished. For another ten minutes I played with a pair of the largest in the group. Were these the elders? They seemed curious about my performance with the dorsal fin. Whenever I put on a good demonstration of swimming and diving there was a noticeable surge of vigour as if dancing through the sea. I had improved my performance greatly since practising with rhythmic music and the more I increased my activity the closer the dolphins came.

I yelled out to Jan to get in. As she got ready two dolphins kept coming over to the boat and looking up at her as they turned on their sides before returning to me. She was just about to enter when they disappeared. We decided to leave as it would not be appropriate to go closer to their feeding pattern in case we upset it.

One week later we met the dolphins again. There were no signs of activity until we were within five kilometres of the Knights, then to the north we sighted gannets wheeling and plummeting. We homed in on dolphin fins and soon they were gambolling around our bow. This time Jan had her first experience of playing with them.

She hurried into her gear and leapt in. They were immediately all around her. She didn't know how many there were but was conscious of a group of five that stayed with her. There were two pairs and a single one and it was always the loner that came closest. She extended her arm and it seemed only a foot away from her outstretched fingers. It wagged its head at her and seemed to want her to dive deep and follow its antics.

She felt they were disappointed when she had to keep breaking off for air and was unable to dive deep, but she cavorted with them for half an hour. Every time she descended they would come rushing up vertically to frolic around her.

These dolphins came of their own accord. The boat was not running to attract them. When they arrived, the music was playing and nobody was in the water. That lone dolphin, White Patch, sticks in Jan's mind. It seemed to want her as a playmate and she will never forget looking into its eye as it made slow, close passes. 'It was a friendly intelligent look; understanding, playful and wise all at once.'

All the time, she was conscious of other dolphins circling at a distance. Towards the end of her dive she played with the dolphins just under the bow of the boat so we could see what was going on. Then she began to feel seasick and it impaired her performance — most frustrating for her. All she wanted to do was forget everything and swim away with the dolphins, copying all their actions.

From that day, Jan was as convinced as me that we should devote all our energy to learning more about dolphins in the wild.

As the boat skittered towards the coast I had time to think things over. It seemed likely we were meeting the same group of dolphins. Close study of individual markings would enable us to recognise each group and to determine whether we were building up a relationship, providing groups were reasonably stable.

The handfins? I had really discovered how to use them trying them out for the first time among dolphins. They control turns to left or right and enable the dolphins to descend or rise.

The fish has a rudder for turns, but the dolphin doesn't need one. His flippers enable him to turn with incredible rapidity about the fulcrum of his dorsal: he points both limbs in the direction he wants to turn and *zowee*, he pivots like a jetboat. His body and tail fluke are transfused with a flowing S-bend of energy, but it is the flippers which initiate the manoeuvre, deflecting him up or down, firing him completely out of the water in a highspeed leap or corkscrewing down into the planktonic dark.

Dolphin Girl Meets Sideband

When Jan first met dolphins while clad in a properly constructed dolphin suit, their responses exceeded anything we had expected. . .

In November 1977, we sighted common dolphins herding baitfish near the Poor Knights. At the outset, flautist Eric Kircher and I approached their vicinity with the inflatable boat and hove to. While Eric sat on the bow playing his flute, a subgroup of dolphins came over and mooched about. From a previous flute session, we knew that when dolphins hover quietly on the surface they are listening to above-water sound. Those of us in the water noticed that we, too, had to remain virtually motionless; any immediate splashing sounds masked the more distant notes of the instrument.

From the ship Jan could see fins crowding around the inflatable. She struggled into the dolphin suit — just in case. I returned to the ship and picked her up because she finds the inflatable easier to work from when she has the suit on and no legs. She lay on the side and rolled gently into the water.

While Eric played his flute a group of about two dozen dolphins circled the dolphin girl, inspecting her curiously for 15 minutes as she dived and curved through the sea. After 20 minutes only three dolphins remained and then, just one. The rest resumed their fish-herding occupation with the main group.

The loner was so curious it came within four feet, spiralling slowly. When Jan descended it dived beside her, coming much closer underwater than on the surface. When she ascended, the dolphin rose too. Every now and then it would disappear but the water was so green and murky with plankton it didn't have to go more than five metres to be beyond her vision. With its sonar the dolphin could easily scan the dolphin girl as it circled in the fog. She knew it was close by as she could hear its sonar whistle calls. Each time she dived, sure enough, into vision it came, circling slowly.

'It was looking at me with that wonderful eye,' Jan told us. She was quite overcome with emotion that a creature like this would come so close, scrutinising her with such interest over and over again. In these circumstances all she could do was to hang there motionless underwater gazing at the dolphin, taking in every part of its handsome, streamlined form so that she might recognise it again.

She called it Sideband; it had a white vertical band on its left side behind the dorsal fin. It wasn't a scar but looked more like a watermark on a painting. A large dolphin, it was a paler grey than most.

Then she remembered to mimic it. The response amazed her. As soon as she began moving like a dolphin, Sideband responded by throwing its tail up and its head down, performing its own exaggerated version almost to the point of absurdity. Jan wanted to laugh for sheer joy.

Topside on the boat, we watched the game with equal delight. You could have bet money on it — every single time Jan rose from a dive, the dolphin would bob up alongside her, gasp and descend. We knew it had to be circling her out in the murk because when she dived it would rejoin her.

After 30 minutes Jan was tired and she rolled into the inflatable, now tethered to the big boat. She sat there alone, stretched along the side. To her delight, right beside her in the water, was Sideband. She extended a hand towards it as the inflatable glided forward under tow.

Her strength recovered when the original dolphin group returned. She slipped in again and dived. There were dolphins everywhere, sweeping beside, under and around her, all defaecating as they passed. She performed a couple of rolls and noticed two dolphins do the same, their white bellies flashing.

After a while the majority withdrew leaving the same trio — Average White, Smallscar and her special friend, Sideband. It swam along slowly with her and when she lost it in the murk, she dived. Each time she did this it would sweep in close and would surface beside her. There was a definite pattern emerging. When she descended it always dived at the same time and came in closer to her than when on the surface.

She was unable to determine its sex because she never caught a glimpse of its underside. It would pass close by her but always on the same level or slightly below. I shot a roll of film of the episode. I was itching to film it underwater but was concerned that this might upset the emerging pattern. Every so often the other dolphins would return for another look.

Twice Jan got out for a rest and each time Sideband swam beside the inflatable. The third time she got back in the water a fourth dolphin joined the trio — Grillmark, because of a grill pattern on its left flank just in front of its dorsal.

By now Jan was beginning to feel exhausted with continual diving and breath holding. She lasted only 10 minutes more, and remembers thinking, 'If I black out now I wonder if this dolphin will come and support me?' She was getting dizzy from staring at the shafts of sunlight shining on the plankton particles, long rays forever angling down into the depths.

The moment Jan decided she would have to get out, Sideband, as if it knew, disappeared.

The total time this dolphin had bonded with her was four hours and twenty minutes. We were sure now that the dolphin suit must have had something to do with this because we had never had such intense interest from them before. In fact, I have not been able to find any other account of a human/dolphin encounter in the open sea of such duration.

Physical Contact — The following day our catamaran met with the same group of common dolphins and two of them deliberately made contact with us as we dangled our limbs among the bowriders. I had noticed that dolphins usually shudder and flee if people touch them unexpectedly, so I suggested we make ourselves accessible and leave it to them. I watched in sheer delight as a dolphin veered and nudged my friend's hand. Just then I felt a dorsal fin glide sensuously between my toes . . .

It became our practice to establish a receptive mood from the outset by playing sweet music through the bows of our catamaran. For some time before entering the water we would lie in the bow hammocks within easy reach of the dolphins. Mutual trust emerged as the vessel drifted to a halt, with the dolphins milling around our bows. At this point we felt we could enter their space . . .

The second time Jan met dolphins while wearing the dolphin suit, to my dismay, one almost succeeded in abducting her . . .

31

Dolphin girl slid out of the net and clung to the leading edge for a while so the dolphins could see her. Two on the starboard bow looked across. She felt the catamaran slow down. She dived. Immediately, five dolphins swam around her, moving slowly and showing great interest. She kept swimming and diving as the dolphins circled, swooping under her, some coming up from behind, some straight towards her or in from the side. She got the feeling she was moving in a set direction but she couldn't tell where, because she didn't want to raise her head from the water and break visual contact with her companions. But she did become aware that the dolphins were quietly leading her in the direction they wanted to go as she sought to stay with them. She forgot about everything and became a dolphin.

Then she noticed that there was only one dolphin left, a large one with a thickness towards the tail that denotes maturity. On the right side of its dorsal were two small scratches. This dolphin kept surfacing beside her, sometimes on her left, sometimes on her right. Each time she caught sight of it again she would dive and the dolphin, beside her, curved its body in a graceful arc and slid below too. Keeping in time with her body waves, the dolphin performed a slow-motion, exaggerated version of the dolphin-kick. Was it mimicking her or giving a lesson? Whatever, this was far from the usual behaviour of a dolphin and it was repeated possibly as many as five or six times. Then it would go on ahead and slowly turn to the right or left to look back at her. She had the feeling it wanted her to follow. Then the dolphin would disappear into the haze only to turn and resume its position in front of her. Sometimes it came up behind her. Once it circled her steadily and she circled too, gazing in admiration at its form. She dived and rolled over. The dolphin didn't roll completely but spun on its side flashing its white belly at her. She was not quick enough to discern its sex. Then it took the lead, waggling its head in that slow-motion dolphin curve again.

Jan noticed that each time it came back, the period of separation was lengthening. During one of these intervals she suddenly realised that she had forgotten the catamaran entirely, because she had been so absorbed with her companion and its antics. She was shocked to find us some 400 metres away and could not believe she had swum so far, oblivious of the boat, her own safety — everything. She saw us preparing to leave in the inflatable. With that momentary glance she felt as though a spell that had been binding her had broken.

The lone dolphin returned three more times and she dived with it repeatedly until it disappeared and didn't return. She wondered if the dolphin, sensing her distraction and concern, had decided to leave her. Its companions were far away.

PLATE 1A

M.V. *Interlock* is a James Wharram design Polynesian catamaran equipped for meeting dolphins. The bow hammocks offer easy water access, and the rondels symbolise the harmony of opposites.

PLATE 1B

The dolphin suit enables more complex responses to dolphin body language. Mutual mimicry and playfulness are primary avenues for communication, as between mother and child.

PLATE 2A

The first time Jan Doak met common dolphins, wearing the dolphin suit, one bonded with her for over four hours – a rare response from this shy and cautious species.

PLATE 2B

Common dolphins, *Delphinus delphis*, have a contrasting black/white hourglass pattern on their sides. They are usually ound further offshore than their larger bottlenose cousins.

PLATE 4A

Not reflections: three *Delphinus* couples (males are beneath) demonstrate love-making.

PLATE 4B

Mother, baby and aunt thirteen months later. The dorsal fins often have a white patch, variable in shape.

The Sound Channel

Many more interlocks with the common dolphins followed, each giving us further insight into body language exchanges which would engage the interest of dolphins and prolong contact. But there was another approach I was eager to test.

Sound is something dolphins and humans both experience, most likely in different ways, but there could be some overlap, some mutual aspect, which might transcend the barriers between us. How would dolphins respond to our music if we played them a special recording that demonstrated in analogue manner, our knowledge of the frequencies they use? Dr Carl Sagan describes how he set up a visual version of this approach to alien minds aboard the deep space probe *Viking*.

The frequencies dolphins employ are four and a half times faster than those used by humans. This corresponds to the difference in the speed at which sound travels through water compared with the air. Normal human speech is at the same frequency as all the mechanical sounds in the ocean. To dolphins we may appear both deaf and dumb. Perhaps if we were to dance to the music they might mimic us, join our game and grasp that we are tuned to the low frequency sound patterns being transmitted.

I needed a special tape-recording on which dolphin sounds were slowed, stage by stage, to our listening speed and frequency, so that they could appreciate what was going on. Then a piece of western music would be recorded in stages to four times the speed of normal sound. This would communicate a pattern at their level of perception.

Both man and dolphin are surrounded by an extraordinary chaos of information noise, thousands of bits of data quite irrelevant to survival. The signal-to-noise ratio is low and it takes complex circuitry for humans to scan incoming signals for pattern. With refined sensory systems, an ability to discriminate between signals and noise becomes increasingly important. Both man and cetacean have evolved as social animals.

In a complex social context generating a heavy static of irrelevant noise, selection favoured the group with the big brain, with its bulky cerebral circuitry, noise limiters, fine tuners and pattern-reading apparatus. But therein may lie the tragedy, for when it comes to interlock between minds in media as different as air and water, both parties have missed the bus. Anyway that was the theory I wished to test.

Then came a meeting with a young Hamilton lawyer, David Harvey. I explained to him that I needed a special tape-recording on which dolphin sounds were slowed to our listening speed, followed by a piece of western music recorded in stages to four times faster than normal.

To my delight a tape cassette arrived in the mailbox from Hamilton. David had used two reel-to-reel recording machines to produce exactly what I had outlined. Furthermore he had put on the remainder of the tape, a long recording of humpback whale songs and the vocalisations of 20 different cetaceans.

At the very end of January 1977 came the most momentous interlock to

33

date, the chance to try the tape with common dolphins.

Halfway to the Knights with an American Bob Feigel, my daughter Karla and her friend Lisa aboard, we met the dolphins. This time there were no birds working. One or two gannets and shearwaters were sitting around on the ocean and occasionally taking off for a few circuits. We had whale sounds playing on the tape and the girls were ringing the cowbell.

The taped dolphin sounds finished and Pink Floyd started. At twice the normal speed 'Echoes' has a crazy carnival rhythm.* Somehow it sounds just right in that setting and the dolphins all around showed clearly they were listening to the tape. When the tempo increased a surge went through them and they began to frolic.

I never thought, when David Harvey and I first tried that tape one winter's day on the coast six months back, that we would manage to play the entire experiment to a group of dolphins and feel certain that they had heard the sequence of acoustic logic it expresses. From the outset David and I had hoped that if dolphins ever heard these experiments, they might realise that the communication barrier between man and dolphin is the great gulf between the wavebands which each species uses. We hoped there might be some significant response if ever they grasped the analogue message on our tape. So I was particularly interested in this quickening of pace with the Pink Floyd piece.

We had been cruising with them in this manner for 25 minutes with Bob imitating the whale sounds on the final part of our message tape. When it concluded I switched off the motor and we glided to a halt on the mirror of calm. Ahead of us the dolphins were milling about at the surface. The only sound was the distant rumble of holiday traffic along the coast. Bob let out another shrill whistle. We heard breathing and splashing from the dolphins. Then it happened. Our ears thrilled to a raspy, hollow whistle — it came from among the dolphins. *One had answered Bob.* I saw a large dolphin surge off to the west. It sounded and emerged, whistled boldly and dived. We were hanging there, Jan, the two girls and me. Bob was standing at the grab rail, whistling across the water and listening to the responses, hardly believing, each time he received a reply, that it was not one of us, but that a sea creature was making such sounds.

Now the dolphins were at quite some distance but we continued to drift silently, listening while Bob kept up his whistling. Then we noticed they were getting closer. They came nosing through the clear water right back to us, defaecating as they came, hurtling under and around the boat. It seemed that Bob's whistle had drawn them to us again. One group stayed close to the boat and others further out.

While we circled, Bob was leaning over the bow, his arm outstretched above the dolphins' backs, whistling to them gently. I saw one answer Bob ten feet from the boat. I actually saw the blowhole moving as it whistled. Bob touched one, lightly rubbing his finger along its side. He said later that he was amazed at the softness of its skin. The dolphin gave a start, splashing water with its tail. It came back up again rolling on its side to look at Bob, then it did a sideways wriggle.

Vocal mimicry is the first stage in communication between mother and child. The dolphins have now shown they can adapt to our low-frequency airborne-sound signals with perfection. Physiologically it is not easy — a highly

---

*'Echoes' was selected because it is an electronic simulation of whale song: mimicry.

consummate feat by an acoustic gymnast. But once the actual production of such sounds was mastered, they rapidly perfected their modelling, just as a man first learning to make a sound on a trumpet progresses to playing a recognisable tune.

In captivity it takes scientists weeks of reward/punishment training to elicit airborne sounds from bottlenose dolphins. But these are lonely, touch-starved creatures with little choice other than to comply with the research pattern imposed on them. In the wild, the dolphins have the trump hand — they can withdraw at will. In this situation man can only make himself available to be taught by the dolphins.

From this point the sound channel seemed as promising as body language as a mode for further exploration and we began to improve our equipment and techniques, year by year.

Interlock with Bottlenose Dolphins

It was not until early 1978 that we began again meeting bottlenose dolphins, *Tursiops truncatus*, as in our initial cavorting, back in April 1975. One of our finest encounters we called *Nudelock Day*. Since it has much in common with later experiences in the Bahamas (Part Three), I will present it here.

Nudelock — 22 January 1979 was one of those diamond days of summer, the calmest and clearest of oceans and a gentle zephyr of wind ghosting RV *Interlock* straight towards the horizon. Among the five people on board were marine scientists Tony and Avril Ayling. Having completed doctorates in undersea ecology, they were on the point of leaving for Australia, where they would begin a programme of Barrier Reef research with Dr Walt Starck, aboard *El Torito*.

It meant a lot to me to be able to show these two 'sea people' the kinds of things that had been happening to Jan and me when we met dolphins with our Interlock setup, and to receive an evaluation of our study methods from two trained scientists.

Tony was involved in the initial interlock back in April 1975, when we danced with *Tursiops* for an hour. Since then, although we had had many intensive meetings with *Delphinus*, we had not had an interlock with a large group of *Tursiops*.

On 8 September 1978, Tony Ayling had seen a group of *Tursiops* playing courtship games and copulating vigorously in Nursery Cove at the Poor Knights Islands. Tony had found it difficult to concentrate on his long-term fish behaviour observations.

A dolphin would swoop above the bubble-kelp forest that fringes the Sand Garden, dive into the top of the fronds, grab a piece in its beak, break it off, swim with it for a while and then drop it. He saw the seaweed game being played several times, on two separate dives.

Now, as he lay on our port bow, Tony thought he saw dolphins leaping a long way to the south. We kept on the same track and they came to us from all over the ocean — bottlenose dolphins leaping. With a gentle southwest breeze, we let the boat steer itself. There were three eager divers in the net with masks on. Frenetic activity began around the bows.

'You can see them coming from away off,' yelled Tony lying in the net, his mask submerged. The water was clear, over 30 metres visibility. The message tape was playing.

Avril overbalanced, fell in front of the net and clung there. My son Brady slowed the boat and Jan and Tony joined Avril in the water. Tony shot a whole roll of film with his Nikonos and fisheye lens.

Jan noticed a dolphin with its dorsal damaged at the base. 'Wade, I think Busy Bee is here!'

As I looked over to the port bow a fin thrust into view — it seemed almost deliberate — and with it was a companion with a short stumpy fin. We got our fin identification file and compared them. It was Busy Bee with its companion, Stumpfin, whom we had met and photographed a year before — on 16 February 1978. Recognising them was like seeing old friends.

Jan put on her dolphin suit. Tony suited up, too. Avril stayed at the bow to maintain contact. She shouted excitedly that one had opened its mouth, showing its teeth as it passed, but not aggressively. Another was hanging vertically in the water, rising slowly as a diver does for a breath. She called it Triplenick.

Jan and Tony re-entered in their wetsuits, the three of them dragging through the water clinging to the bow net, with dolphins converging from all sides — up underneath and from behind — to inspect the humans, dangling from the orange net. The dolphin movements got slower and slower, until some were almost motionless around the bows.

Triplenick became excited, dashing about as all three divers swam away from the catamaran. Jan saw a dolphin with a piece of fishing line hanging from its tail. Three times she yelled as the dolphins moved slowly around them close to the surface. She noticed one to the left look directly towards her and let out a huge volley of bubbles like the exhaust from a scuba regulator. It seemed to mimic her action, but not the sound.

Each time the three dived, the dolphins dived also and swept in close, head-on. Or they would come up behind and curve around in front, only a metre away. The closest they came was when they were moving in from behind. Tony said later, 'They came closer than I've ever experienced before — two feet from my body, six inches from my hand.'

And Jan, 'They were huge, such massive, strong forms moving in front of my eyes. I've been used to swimming with the smaller pelagic dolphins, *Delphinus*. These *Tursiops* were giants in comparison.'

Triplenick kept paying Jan, in the dolphin suit, a lot of attention. It would sweep around her, exaggerating the dolphin-swim and tossing its head playfully like an exuberant puppy. When she dived it came in head-on, and as she finned horizontally, it swam beside her, eyeing her intently. When she rose for a breath, to her amazement it stood vertically in the water and rose up to the surface watching her as it went. It was so comical that she laughed. She had the urge to throw her arms around it in a hug. The gesture was repeated several times.

She found later that Tony and Avril saw the same gesture but didn't identify it as Triplenick, even though this dolphin appears in several of Tony's pictures. Twice Tony watched a dolphin hang motionless, upright in the water, just below the surface and then sink slowly, tail first to about five metres, emitting a small trickle of bubbles. On one occasion it made a slow, yawning gesture. Another dolphin approached him horizontally and hung motionless three metres away looking at him. In 15 different encounters with *Tursiops* he had never seen such behaviours.

Jan moved closer to the other divers to see what responses they were having. A large dolphin dived towards her and turned broadside on. Passing two metres away, it slowly opened its mouth twice, open and shut, open and shut. She could see its sharp teeth but this did not appear to be a threatening gesture to her. For the first time she was able to recognise one dolphin as a male. She was three metres below when he swam past her, up-ended before descending with his underside towards her. Clearly she saw the two 'in-line' slits — no mammaries.

Six metres away she saw Avril and Tony together, with Triplenick circling them, before rushing over to her and then back around the others again, still doing the exaggerated dolphin-swim and tossing its head. She was reminded of a high-spirited teenager. Then she heard the message tape playing. She was near the starboard hull. Whale sounds were floating out and she noticed three dolphins quite close to the hull in a stalled position, just hanging there, flippers drooping, slightly hunched as if listening to the sounds.

During this time I was operating the hydrophone and the message tape, shooting movie and still film, as well as assisting Brady with the ship. Although I was bursting to get in with the dolphins, it seemed most important that the two biologists should have the fullest possible experience of our setup before they went overseas. The message tape was conveying an analogue expression of the frequency we use, paralleled with low frequency whale sounds.

I heard a dolphin make a blowhole sound in air as it breached by the bow and surged off at right angles — a loose 'raspberry' sound. Gradually the number of dolphins around the catamaran diminished until only six remained, including Triplenick — two sets of three.

I wanted to film them at close quarters but the cat was drifting south and the dolphins stayed put. As Jan swam to the boat it may have seemed as though she was leaving — she was feeling seasick and tired. Perhaps they sensed this. We were all hungry and in need of a break. Towards the end of this interlock two dolphins came curving in on their sides and then made a series of vigorous tail slaps. I replied. We had been together continuously for one-and-a-half hours, our longest interlock ever.

We lunched, drew fin and tail identifications and sailed on a northerly tack parallel to the coast. As we turned east, we saw a pair of dolphins leap high out of the water forming a neat pattern, like our sail symbol. They did this three times and were about 500 metres away. Then at a quarter to three, several dolphins joined the boat. One of them was Triplenick. This easily recognisable dolphin made three approaches to the starboard bow and then headed off, at right angles, to the southeast. I took photos of the fin. Then, following in that direction because it seemed like a gesture, we saw lots of dolphins approaching.

The second interlock began at five past three and lasted an hour. We were about five kilometres off Tutukaka, in line with the Pinnacles. This time Jan, Avril and Tony were nude. There were about 50 dolphins including several youngsters, one so tiny it must have been recently born. About 450 mm long, it still had vertical stripes (birth folds) on its sides. Its mother and another dolphin kept it between them.

This time I just *had* to get in. The ocean was warm and deep blue — 30 metres of clear vision. I dived down with my 24-millimetre lens, dolphin-swimming, and found I could get 30 dolphins in my viewfinder at a time. Using High Speed Ektachrome, I took some shots at 1/125-f5.6, then 1/250-f4, in case of excessive movement. Jan and Avril looked exquisite with their long hair flowing in rhythm with their dolphin movements.

Afterwards, Avril sent us a painting she had done to celebrate her first dolphin experience — two nude human figures and five nude dolphins. With it she wrote: 'Suddenly there were dolphins all around, big bottlenose dolphins, arriving and flying through the rays of light that shot the deep blue sea as if it were silk. There were perhaps 60 dolphins and we felt beckoned to be in there with them. With our bodies naked and trembling with excitement we repeated their swift movements; effortlessly Jan and I flew with our dolphin-kicks, looking down into the cones of light that swept from our bodies into the darkness below and then around at our partners in this ballet.'

Jan felt that the dolphins were scanning her with their sonar at a range of five metres and then moving in close for visual scrutiny. 'I felt so free and everything seemed right. We were in our natural state just as they were.'

She counted 17 and then lost track. There were babies of all sizes, then others making up, as it were, the whole village progeny. She saw one huge old dolphin, very dark with mottled grey blotches, and one peculiar individual with a 'punched' snout like a bulldog.

Tony and Avril were curving in unison like dolphins. Jan and I joined them, in perhaps the most sublime moment in all our dolphin days. Avril and Tony held hands and caressed to show our touch responses. The dolphins did likewise, rubbing along each other's sides and folding across one another. Trustingly, these dolphins had brought their young to see us and were allowing them to come close.

When three divers were below at once, each was approached in turn, the dolphin heads swivelling to and fro as they scanned the human forms. We are so limited underwater, unable to communicate, completely cut off in spheres of 'self'.

Jan noticed the tiniest baby swim up under its mother and nuzzle her underside to nurse. Meanwhile another dolphin in the role of 'aunty' came up under the baby, gently rubbing it and holding it securely between the mother and itself. Then, as the baby resumed its position beside the mother, it was flanked by the other dolphin, the two adults rubbing their sides against the baby, as if cuddling it. I took a picture of them. I am reluctant to use my camera when we meet dolphins, as I feel like a nosey tourist in a village. But with the second meeting, I felt we were accepted. The dolphin pictures Tony and I took that day are the best I've seen of bottlenose in the open ocean.

Seaweed: I noticed a dark dolphin with a strand of worn-looking bubble kelp wrapped round its forehead. It came past me again on the same level and course, with the kelp around one flipper. Each time, the pattern of approach, course, level and distance were the same. A gesture?

Jan saw the same dolphin earlier with the kelp draped around the dorsal and later, around its caudal peduncle. Tony saw it with the seaweed in its mouth when down deep, and around the tail fluke when near the surface. But *nobody* saw it change the position of the seaweed.

By this time I was alone in the water. Out from the midst of the dolphins came a strange shape — a large bronze whaler shark. My first reaction — is there any danger? I looked at the dolphins cavorting with their young. They showed no fear. A calm feeling flowed into me. I took a shot of the shark as it cruised past, eyeing me as it swam to the southeast and away from us all.*

My film was finished. I gave Brady the camera and got a beach ball, tossing it to Tony over the water. But the ball is a surface thing and we needed something more submersible, a slow-rising, buoyant thing like the seaweed — a bit of nylon rope perhaps, to start an exchange game. While Tony and I were tossing the ball we didn't notice, Jan said later, a dolphin leap out alongside us.

*I have also seen a large hammerhead shark swimming with common dolphins who showed no concern; but I have an anecdote of bottlenose dolphins attacking a mako, killing it in 15 minutes and tossing its body clear of the water.

In our second interlock the dolphins moved in large groups rather than as individuals, and frequently dived deep. Again it was the humans who withdrew from exhaustion. The day, in total, provided the longest period of intensive interlock we had ever experienced — Avril's first and Tony's best, out of about 15 contacts.

We got home at five o'clock, under sail, in time for the Aylings to catch the bus on their journey to Australia. Avril was ecstatic, Tony no less.

Project Interlock Extends

In April 1979, we took a further step in our research. So far we were satisfied that it was possible to gain acceptance by two species of dolphin, if they were treated in a communicative manner. To provide a context for our experiences, we needed to know more about what happened when other divers met dolphins, and whether our methods would improve the frequency and quality of diver/dolphin encounters. At heart we felt that until dolphins met with more consistent warmth and receptivity from our own species, we could not expect to get much further ahead.

I had published *DIVE* magazine for 15 years and gained a pretty good measure of what was commonplace or rare in the diving world. I was confident that diver/dolphin contacts were, for the most part, brief — hello-goodbye encounters with quite a lot of fear on the human side, because people weren't quite sure of what to expect.

So in 1979 we began publishing Project Interlock newsletters in New Zealand, Australian, American and British diving and boating magazines, outlining the procedures we use to establish friendly rapport when meeting dolphins. A questionnaire was devised to standardise information received, and written narratives were also encouraged for publication in subsequent newsletters.

From the outset we were most anxious to promote an ethical code for interspecies encounters. All of the finest experiences on our file show that people should not impose themselves on dolphins — they gain little.

Photographers, eager to get underwater shots, have been known to chase after dolphins in boats, leap in front of them, filch a few fleeting glimpses and then resume pursuit. One author tried this technique on Hawaiian roughtooth dolphins and provoked them to aggressive displays. No photos were obtained. The finest episodes with dolphins occur spontaneously. The dolphins make the choice. If humans are unafraid, an encounter may intensify.

Our newsletters stated explicitly that if dolphins, orca or whales are chased, encircled or buzzed by outboards, they will withdraw.

Meeting Dolphins — First we described our own setup, the catamaran and its equipment, outlining briefly some of our experience in gaining the dolphin's acceptance. In essence, we stressed the need for a sensitive approach involving mutual respect, and a creative mood such as music may induce. We urged that harassing manoeuvres or any forcing of human presence on dolphins should be totally avoided. Then we outlined a number of friendly gestures we found successful in communicating our intentions to dolphins and engaging them in gamesplay.

Amongst a group of dolphins around the bows there may be certain individuals that are most attentive and keep returning again and again. We are quite sure that dolphins have varying personalities or behavioural traits, despite the uniformity of their bodies. It is important to observe these individuals and respond appropriately as you learn to recognise them from distinctive fin markings, body scars and other distinguishing features. You may then find you meet the same dolphin on future occasions and can respond appropriately.

In this way a body of shared experience may develop into friendship. A camera with high-speed film is useful for recording fin patterns. We use a cardboard dorsal fin as a template on the outline of which we record distinctive nicks and notches with dates and locations. Scratches and superficial marks may last for only a few months. This material is presented in Appendix A.

Feedback — Feedback from Project Interlock newsletters was quite staggering. As accounts were published of humans meeting dolphins, people sought similar experiences and reported them to us. A sequence of reports encircled the New Zealand coastline from Spirits Bay to Foveaux Strait. After that they started extending across species and around the globe until we were filing interlocks with bottlenose, common, dusky and Hector's dolphins; followed by spinners and spotted dolphins of tropical seas.

By early 1980, we began to notice certain patterns in these reports. Duration seemed to be the simplest way to evaluate human/dolphin encounters. It became apparent that the more interesting and responsive the divers were, the longer the dolphins would remain. Divers who swam directly towards dolphins, or attempted to grab them, were left staring into the blue. It would be impossible for scientists to study dolphins passively in such situations; they *have* to become participants, often in conflict with their own disciplines.

The most interesting interlocks were those we termed 'open-ended', where the people involved had to withdraw because of exhaustion or some other factor, but where the dolphins were prepared to continue the encounter even, in one case, pulling at a diver's fin as he left the water.

We learned that human fear was a major factor in cases where dolphin approaches had been brief. While sometimes loathe to admit it, many people have a certain distrust towards sharing space with a creature larger and faster than themselves — something shark-like and unpredictable.

An amusing example took place off Mahia Peninsula on the east coast of the North Island, New Zealand, in November 1979. A novice diver was hovering below the boat while his companions were having lunch. They saw three common dolphins detach and head towards the diver, each encircling him once, at high speed, before rejoining the main group. And then, a great cloud of bubbles arose, and the diver emerged like a submarine missile screaming obscenities . . .

As our newletters presented dozens of satisfactory accounts, fear diminished and motivation was heightened. Many experienced divers described an encounter with dolphins as the highlight of their underwater careers.

Whales Alive Conference

In 1983, for the second time in my life, I was invited to attend a global conference on cetaceans. This time the venue was Boston, Massachusetts; the theme was 'Whales Alive — Non-consumptive Utilisation of Cetacean Resources', which strikes a strangely anthropocentric note, but this was a pragmatic bid by conservationists to convince our hard-nosed world that there *are* worthwhile alternatives to killing cetaceans.

One of the papers I submitted was for the session entitled 'Benign Research with Cetaceans'. In its revised and updated form, I offer it as Appendix B. It is a portrait of human/dolphin encounters parallel to our own experiences, and provides a normative context for the rather extraordinary exchanges I describe later in this book.

We found many instances where other divers experienced the same signals we had encountered, but in many cases, lacking the context of experience, they did not grasp the significance of gestures such as seaweed gamesplay, bubble gulps, signal leaping, sexual behaviours and such like. Getting no response, the dolphins did not intensify the exchanges.

To conclude this section, an episode involving a lone dolphin and two scuba divers serves as a bridge between our meetings with social groups and the history of solitary dolphin encounters that follows.

After ten years' research our understanding of solitary dolphin behaviour has deepened as much as any simple explanation of the phenomenon has eluded us. For prolonged periods individual dolphins have shown much more complex and intimate responses towards humans than are generally manifested in the group situations of this chapter. But the phenomenon cannot be neatly categorised; there is a continuum from one aspect to the other. Occasionally a lone dolphin has shown curious and intensive responses to divers but these have not been followed by further episodes. One can only surmise that the individual may have resumed its social group — or possibly, died.

Near Moeraki Point, South Island, New Zealand on 15 March 1981 a bottlenose dolphin circled a dive boat. For 30 minutes P. Winders snorkelled playfully with it before he and his companion donned scuba gear and descended the anchor line in search of rock lobster: 'The bottom was about 15 metres below. At the sand we started our way along the drop-off. Next thing the dolphin was there, suspended tail uppermost, head nearly on the sand. It remained motionless for several seconds at a time. It did this repeatedly. At one stage I was poking around in a crevice when I had the feeling I was being watched. I looked around to find the dolphin's head about two feet from my own, trying to see what I was looking at.'

When they surfaced, the dolphin accompanied their boat for the full half-hour journey home, following them right through the channel into the shallow surf.

Stepping back from it all, there seems to be some potential for a special relationship with an individual, free-living dolphin, based on mutual curiosity and respect. Certain subtle messages and consistent patterns of behaviour keep appearing that almost defy rational analysis. Perhaps our intuitive, pattern-recognising processes need to be attuned to grasp the significance of such gestures and so, respond spontaneously. This may lead to the etiquette required by each species for play/touch communication as a basis for interspecies trust and exploration. Who knows just how complex the relationship between terrestrial and ocean minds might become?

Dr Fred Donaldson writes: 'Play has a survival value much broader in scope than we have imagined. This is not the survival value of man in contest with his fellow creatures; it is the survival of man in kinship with his fellow creatures.'*

*'Play & Contest in Human/Animal Relationships', F. Donaldson, Ph.D, Lomi School. Bull. Fall, 1972.

Appendix A: Meeting Dolphins

It is difficult to say which ingredients in our approach are essential, but we have had many successful, prolonged interlocks. Most people find that while dolphins may bow-ride, they leave if you jump in with them. These dolphins may be frightened or holding out to establish more equitable human/dolphin rules.

'We feel it is vital to avoid any tendency towards manipulating dolphins to our own ends. We must not harass them, encircle or in any way force our presence on them; remember the *Marine Mammals Protection Act*. Even using a camera should be done with discretion — it forces you to behave in an active manipulative way and can spoil the spontaneity of interlock. A receptive mood is best for all on board. It is better to leave cameras out of it for a while or dolphins will find you dull — it's hard to be playful and spontaneous with a handful of 'f' stops.

'When we sight dolphins we just heave to or hover quietly about in their vicinity. From our forward beam hangs a bell which we ring about six times when they surface for air. This establishes our acoustic identity. If they are not too busy herding fish, they will leap out and head over to us. We play carefully chosen music through our bows, especially flute and wind instruments. This tunes us in to their presence as they bow-ride.

'In the first stage of interlock we let them see us as we lie on the bows of the catamaran or in the hammocks suspended between. Often they turn on their sides to scrutinise us. Mutual trust is being established. As we make ourselves accessible to the dolphins, each species is equally vulnerable to the other. At this stage it seems wrong to try to touch them — a sudden lunge may succeed but that is really rape and startles the creatures. The interlock approach is to hold out your hand near the water — show them a human limb, its joints and expressiveness, demonstrating this unusual appendage to a curious alien. The dolphins may examine it, swimming on their sides, and eventually one may approach and nudge your hand.

'With powerboats, bow-riding is potentially dangerous and a safety harness may be necessary. That is why we prefer sail. We know they're not attracted merely by engine noise. Most of our best interlocks have occurred when it was so calm there could not have been any appreciable pressure waves for the dolphins to ride. Despite what cynics may say, dolphins have definitely shown that it was curiosity that motivated their approaches to us in such circumstances.

'Once mutual trust has been established you can slow the boat while a pair of snorkel divers enter the water and engage their interest, dolphin-diving down and staying below as much as possible, turn by turn. Meanwhile the boat may hover nearby, the motor stopped. If the divers are frolicsome the dolphins may stay around and even teach them tricks. It is a mistake to keep swimming directly at them or constantly aiming a camera lens. This may appear intrusive and domineering — it certainly isn't communicative. We have found it best to extend to dolphins similar courtesies and thoughtfulness as when visiting people in a strange village.

'In Polynesia I have seen men land from a yacht and walk among the huts pointing movie cameras at every bare-breasted maiden they saw. The Islanders remained stoically silent and unwelcoming. Unless we respond to dolphins in a communicative, joyful manner, we appear boring to them.

'We have devised a range of activities that heighten interest in us. Our dolphin suit and other gear are not essential, but they seem to signal our intentions effectively. I would argue that it is not anthropomorphising to treat dolphins like villagers with an utterly alien culture. Many animals are curious to explore humans if we meet them in a humble and receptive manner. Like Island children, some dolphins may start to mimic and seemingly lampoon your gross attempts to mimic them. This is the beginning of body language communication, much the same as between mother and child or people from different cultures. Try to show feeling and they may respond. Suppress your intellect and behave spontaneously. Listen for their signature whistles and talk back through your snorkel. Hold your partner as you swim, exchanging sounds with each other and interacting inventively. Try moving with the dolphins as they circle, anticipating them, watching keenly for small signals you can reply to with feeling. At some stage you may find the action slows, with quieter movements and penetrating eye contact, as a more subtle bond develops with certain individuals. At such times a dolphin may brush or nudge you gently. *That is no accident.*

'For the most intimate interlocks it is best if no more than two people enter the water. If you see a close bond developing with another diver be careful not to interrupt. A large melee of dolphins and divers can be similar to a cocktail party and just as superficial.

'Don't be too disappointed if interlock is not successful. Dolphins may have a variety of reasons for not playing — an important fish-herding manoeuvre may be in progress. Such warm-blooded creatures have a high calorific demand, and maintaining adequate food supply could be difficult at times in an over-fished world. So if it doesn't happen, don't persist. Leave it for another occasion. In some cases, because you signalled your intention, the dolphins may return later in a playful mood, once work is over for the day.

'We have seen situations where a large group of common dolphins were engaged in fish-herding manoeuvres yet a sub-group was still ready to detach itself and play with divers provided the boat hovered quietly in the vicinity. Dolphins that have never encountered divers before can be expected to be rather shy.

'Encounters with dolphins in the wild are a rare and precious experience, considering that they share the globe with five billion of us. People who have had such fortunate encounters should write down all that can be recalled while it is fresh in the mind. It is best to make notes as soon as you leave the water, interviewing companions for their points of view. In years to come you may not believe what actually happened, unless you have made a good record.'

Interlock Questionnaire

Sea & weather conditions:
Date & time of day:
Boat name:
Position:
How did boat/dolphins first approach?
How did dolphins behave *before* you got in?
Number of divers in water?
Number with scuba?
Person first in?
Dolphin species?
Approx. number of dolphins?
Any with peculiar markings?
Were there any young present?
How small?
Any surface or U/W shots showing recognisable fins?
What do you think held their interest so long?
How many cameras in water?
Owners' names and addresses:
Any unusual items of dolphin behaviour such as:
— tail first sinking?
— jaws opening and closing?
— seaweed trailing from fins, jaws etc?
— bubbles gushing from blowhole?
— defaecating as a possible signal (ie close to a diver and right in his field of vision)?
How did the divers behave such as:
— snorkelling down frequently?
— swimming coordinately in pairs etc?
— forward rolls?
How close did they approach divers at any time?
How did interlock terminate: did the divers/dolphins withdraw first?
Duration of interlock (approx):
Any special remarks:
Please add your name and address and send this report to
Project Interlock
HQ, Box 20,
WHANGAREI, NEW ZEALAND.

Appendix B: Anecdotal Reporting of Human/ Dolphin Interaction

An overview of Project Interlock files shows several recurrent patterns in the responses of dolphins when they encounter friendly humans. Some responses extend across several species; others are species-specific. Certain behaviours may well be intraspecific (such as copulation, feeding and manipulative play) while others seem to be responses to the observer. But, with large-brained mammals such as these, the very presence of people makes it difficult to draw a firm line. On two occasions, for example, three pairs of common dolphins seem to have made a deliberate demonstration of copulation for the benefit of human observers, either in unison on the bow of a boat or while circling a scuba diver.

To swim in the sea and make an objective study of dolphins may be near impossible but we can learn something of their capacities from the ways in which they have behaved towards participant observers.

I claim that the collection of such anecdotal material is a valid avenue for benign cetacean research. While the non-scientific observer may not always be reliable, if certain behaviour patterns *are* actually present in a particular species, or across species, these will become so abundantly obvious when sufficient observations are amassed as to make observer error unimportant. At least they will suggest where the trained observer might direct attention, given that ocean field study is so difficult. Considering the flexibility of dolphin behaviour, any recurrent responses towards humans must be of significance.

I will now review some of these with sample quotes from observers. It should be understood that such examples are offered as typical of many similar episodes on file, including our own experiences.

When dolphins are first sighted by the diver, one of the most frequently reported behaviours, even in encounters of short duration is the *head-on approach.*

Dave Ellery, Bay of Islands, January 1981: 'I dived just under the surface to meet the two dolphins (bottlenose) heading straight for me. One of them was pretty big compared to the other. They swam in circles for a few moments, then disappeared from sight. Suddenly the two came straight for me from behind.'

John Crossman, Rabbit Island, 30 October 1982: 'On a number of occasions the dolphins (common) approached us head-on, either diving under or passing over at the last minute, often finishing a pass by jumping through the surface.'

Jeff Couchman, Bay of Islands, August 1982: 'Again I entered the water by myself, to be confronted by the four (bottlenose) dolphins closely grouped on or just under the surface, making their usual head-on pass. I swam with them for about 15 minutes, during which they would come straight at me, just beneath the surface, pass very close, dive straight under me to 40 feet and then disappear for a few minutes before repeating the performance.'

PLATE 5A
Interlock begins with five bottlenose dolphins, *Tursiops truncatus*, leaping as rhythmic music is transmitted underwater from R.V. *Interlock*'s special sound array.

PLATE 5B

Credit: Barry Fenn

a perfect day passengers from the scow *Te Aroha* meet bottlenose off Great Barrier Island, New Zealand.

PLATE 6A
Nudelock Day: our most intensive encounter with a tribe of bottlenose dolphins. Among them were a number of individuals from previous meetings who seem to have signalled to us when we were slow to recognise them.

PLATE 6B

PLATE 7A

n 'the swirling' eighteen dolphins performed complex manoeuvres.

ATE 7B

PLATE 8A
Eventually mothers and babies came close, and the tiny dolphin between two adults 'nursed' just after they pass and turn (below). Dolphin babies feed every fifteen minutes day and night.

PLATE 8B

The above were snorkel divers but Matt Conmee was using an electric 'towpedo' and miniature scuba at Poor Knights Islands, 21 August 1982: 'Groups of three, four and five (bottlenose) were swimming towards me on a collision course, in line, abreast formation and then turning together at the last moment, always, it seemed, with the smallest of the group closest to me, as I turned to match a parallel course. It never occurred to me at the time, but in this way the dolphins could have steered me anywhere they desired.'

Certain comments suggest these 'head-on' approaches may relate to the dolphins' initial sonar scrutiny of the diver.

Dave Collins, Deep Water Cove, 21 March 1981: 'I'm resting on the surface when a single dolphin (bottlenose) in a group of 15 approaches rapidly head-on. I dive towards it on impulse and am surprised that the dolphin has done the same thing. We don't collide because it glides effortlessly beneath me. Then a dolphin moves in slowly looking at me, its head swivelling up and down three times, emitting clicks madly. I have the positive feeling I've been fully computerised within a second of time. I dive down. The dolphin follows and wants me to swim in a circle to the right again . . .'

Simon Brown, Cavalli Islands, 13 March 1984: 'Once a large adult (bottlenose) turned back and, from a range of about two metres, examined me. I could feel the clicks almost coming from inside my head. I'm not sure I liked it much.'*

In particular, where there is more than one observer, it may be noticed that *lone* dolphins will approach divers closest from behind.

Mike Oliver, Poor Knights Islands, 21 August 1982: 'I noticed the odd lone dolphin chasing solitary divers and veering off when the divers realised they were being followed and turned around.'

Dave Campbell, Matai Bay, 1 January 1982: 'At first the dolphins (bottlenose) stayed at the limit of visibility (about six metres) but soon came much closer — to within a metre. We dived to follow them as they swam below us and they generally led us around in circles, just keeping out of arm's reach. Once, as I dived down, one dolphin passed very close behind me and bowled me slightly in its wake.'

Mark Carrell, Cuvier Island, 30 March 1982: 'I heard a squeal behind me and turned to come face to face with a large dolphin (bottlenose) hovering in the water about four feet away as if he wanted to play tag.'

Snorkel divers often remark that dolphins came closer and showed more interest while they were below, withdrawing further when the divers surfaced.

Les Grey, Owhiro Bay, 7 February 1979: 'The dolphins (common) didn't pay much atention to me when on the surface — but as soon as I dived down and dolphin-kicked they would "buzz" me like jet planes in tight formation to regroup.'

Jim Nollman, Careyes, Mexico, 21 December 1982: 'Richard comments that the dolphins (spinner) swim right up to him when he is completely submerged, sometimes upside down. But when he surfaces they double their distance from him.'

Mike Young was scuba diving at Kapiti Island, 6 March 1981, while his

*A scuba diver who feigned death experienced real discomfort as bottlenose dolphins buzzed his prostrate form.

companion was 60 feet above on snorkel: 'The common dolphins came rushing straight for me in groups of five or six. They would take a good look as they cruised around me within two metres and move off to the surface towards Bruce on snorkel. They circled him before moving well away from us both for a blow and a breath. I feel this action, going some distance from us to breathe, is quite significant. It would have been quicker for them to have surfaced near Bruce.'

Poor visibility has a limiting influence on the development of human/dolphin encounters. In many cases human motivation is an inhibiting factor, it not being appreciated that dolphins have an acoustic reality, able to maintain contact with the diver when he cannot see them. Once this is understood by the divers, motivation increases and prolonged interlocks can occur in poor conditions. (In six inches' visibility Horace, the lone bottlenose at Napier in 1978, pulled each of Quentin Bennett's swim fins, thus parting his legs. He then slipped between and carried the diver across the surface on his back.)

John Crossman, Mt Maunganui, 7 June 1982: 'I joined the school (about 20 common dolphins) on snorkel by intercepting their course. On making contact each member passed beneath or close by, rolling on its side to investigate me before passing into the gloom (visibility 15 feet). This initial contact occurred in shallow water. Although I was close enough to have touched several as they passed, there was no indication on their part of any real interest in me, (sic) apart from the initial investigation. Visibility was so poor it was impossible to establish their behaviour or level of interest beyond 15 feet.'

The duration of interlock (human/dolphin encounter) may be extended by communicative responses and creative play on either side. Much of our own work has involved intensive study of this aspect, using both body language and acoustic modes. The simplest form of body language is mimicry of the dolphin swimming pattern.

John Montgomery, Marlborough, 9 May 1981: 'I had the same two animals (Hector's dolphins) swim with me, within inches at times, for about 30 minutes. The only problem was that I could not keep up. They solved this by coming back for me. I had very close contact with these two. At no time did I try to touch either of them and they stayed about a foot away, one on each side, doing barrel rolls and mimicking me, or was it me mimicking them?' (He was doing a dolphin-kick.)

Mark Carrell, Cuvier Island, 3 March 1982: 'They (bottlenose) came closer when I was doing the dolphin-kick and actually seemed to be swimming beside me to teach me how to do it properly.'

Reg Lawson, Little Barrier Island, 24 February 1980: 'A large (bottlenose) dolphin swam alongside one of our divers who was doing the dolphin-kick. It had a large stomach like his own and may have been a pregnant female.'

Matt Conmee was mimicking dolphins (bottlenose) using an electric 'towpedo', Poor Knights Islands, 21 August 1982: 'Frequently, when on a parallel course I would do one or two barrel rolls with no obvious reaction. Allan told me afterward that the dolphins immediately behind me were mimicking my barrel rolls.'

Part Two details how Horace, lone dolphin at Napier, exhibited a capacity for *mirror image mimicry*, which seems to indicate a capacity for symbolic thought.

There have been a number of reports, some documented with photos, of dolphins indulging in complex games with pieces of seaweed, either intraspecifically or in reciprocal games involving humans. At times this seems to be linked to a grooming, auto-stimulation behaviour involving rubbing and writhing amidst kelp-covered rocks.

Project Interlock newsletters report groups of bottlenose dolphins tossing seaweed in the air and keeping it up there among the group as in a beach ball game — tossing a piece, leaping and landing on top of it, swimming with a piece draped on an appendage and then switching it to another, and another.

Jeff Couchman, Bay of Islands, August 1982: 'I was then surprised to see one of them (bottlenose) repeatedly lunging half out of the water and flicking seaweed into the air five or six times, as if playing.'

Jill Smith, Kaikoura, 5 April 1981: 'One dolphin (dusky) had a piece of weed on its dorsal fin and there were two in hot pursuit as though the object of the game was to get the weed off it.'

Matt Conmee, Poor Knights Islands, 21 August 1982: 'A very large bottlenose swam at a pace I could easily match with the aid of the "towpedo" and it led me into the quiet shallows where I saw the other dolphins down on the reef and against the cliff wall, muzzling through the *Ecklonia* kelp, much as a dog in long grass, rolling on their backs and hanging inverted in the weed.'

John Crossman, Mt Maunganui, 30 October 1982: 'One of three loose groups of bottlenose dolphins congregated near a rocky outcrop on the end of a reef and made continuous passes through a narrow gap between two exposed rocks. Others rolled about on the bottom apparently rubbing against the weed-covered rocks.'

We have an account of orca behaving similarly at the Rubbing Rocks, British Colombia (Prof Pat Hindley).

John Crossman met seven bottlenose in a shallow, confined rocky inlet near Whangaroa Harbour, 12 January 1985: 'They appeared to be rubbing, certainly sliding over and between weed-covered rocks. The group acknowledged my presence by each making a pass and having a look at me.'

There are many ways in which dolphins make what seem to be signal gestures to people such as: (1) leaps at significant moments, ie just on leaving; (2) deliberately splashing people on boats after close eye contact; (3) vocalisation and whistling through the blowhole, after people have made communicative calls or whistles; (4) repeated short swimming spurts in a certain direction, which we term 'leading behaviour', to guide a diver to a lost partner or a lost object, or to lead a yacht past an obstacle; (5) defaecation; and (6) touch, the most ultimate of gestures.

(1) John Crossman, Mt Maunganui, 6 June 1982 (after interlock with common dolphins): 'They put on a display of jumping and tail standing in the harbour entrance before continuing on towards Karewa Island.'

(2) Bob Tillet, Northland, 2 January 1982: 'It (bottlenose) splashed its tail towards the young people in the bow while they were trying to take photos and their cameras were wetted by the sideways movement, which was quite deliberate.'

(3) Margaret Jongejans, Northland, 8 August 1982: 'Approximately 4-6 dolphins (common) were with me at the bow. Right from the beginning of

the interlock I would call intermittently "te puhi"* and soon I noticed they would surface and make a very exaggerated breathing sound like "poof". It was definitely a response. I trailed my right hand in the water quietly for a while. They consistently swam very close to it. Near the end of the interlock one of the larger ones raised its tail after a breath and *touched my hand*.'

(4) Terry Harris, Hen and Chicken Islands, 1982 (separated from his diving partner): 'Four or five (bottlenose) dolphins were alternately heading in a beeline to the shore, then straight back to me. I got the distinct impression that they were advising me to go in that direction. Upon their departure I decided to swim in the direction they had been "showing" me. After about 40 metres, I found my partner who, it later transpired, did *not* see the dolphins.' (In Part Two three episodes are described in which the lone dolphin 'Donald', in the UK, led one diver to a lost camera, another to a lost scubatank boot, and escorted two scuba divers separately through murk to their anchor line.)

(5) Defaecation as a signal was slow to be recognised by us initially. Gradually I came to suspect that this was a deliberate gesture rather than a random body function. I wrote about it in newsletters and included it in our questionnaire. To my surprise we received a great many reports which corroborate and extend this observation. Often, with bottlenose dolphins, snapper were seen feeding on the excreta. (It is quite common for reef fishes to eat excreta of other fishes.)

Mark Carrell, Cuvier Island, 30 March 1982: 'We swam with them (bottlenose) for about three-quarters of an hour. They defaecated close to us in a gesture. A school of snapper showed no fear and attacked the faeces with relish.'

Matt Conmee, Poor Knights Islands, 21 August 1982: 'Suddenly I was enveloped in a mass of swirling, undulating, blue-grey shapes, excreting faeces excitedly before me and emitting shrill whistles that reverberated through my body. Huge snapper milled about below the dolphins (bottlenose), darting in to feed on fresh excreta.'

Kevin Hope, Poor Knights Islands, 13 September 1982: 'My first encounter was two larger dolphins (bottlenose) swimming towards me. Then they turned. On doing this they excreted faeces in large amounts in front of me. Snapper were feeding on the faeces.'

The significance of this gesture is open to conjecture, but in the total context no diver has ever mentioned feeling it was a threat. Some have felt it signified acceptance. (I note that sheep and goats often urinate when meeting people, as if providing a chemical greeting.)

When examining my pictures closely some months after an encounter, I found that a bottlenose dolphin I had photographed because it was making the faecal gesture, was the very same one I had met on a previous occasion, but I was not aware of this as I took my photo. Had I been more perceptive I would have responded as though it were an acquaintance.

(6) Many divers make fruitless efforts to touch dolphins and there is much evidence that dolphins prefer to make this decision themselves once a trustful relationship has been established — as with Margaret Jongejans above. Our files have accounts such as the beach strollers who spontaneously called to a pair of passing bottlenose dolphins, which came in twice around their legs to be stroked (Ray Gardner, 19 December 1980).

Another bottlenose approached a father and daughter in a canoe, spy-hopping

*Analogue Maori word for the dolphin's breathing cycle.

alongside to be patted (Ken Volkner, 2 February 1982).

During his second interlock at Mt Maunganui (30 October 1982), which lasted two hours, John Crossman records: 'The big dolphin (common) made a close pass above me before assuming a hunch-shouldered, tail-down, stationary posture. I was able to reach out and touch the tail edge — not unlike a well-worn mudflap. I was somewhat tentative about such contact, particularly with that powerful-looking tail. Its response, however, was simply to glide smoothly and almost imperceptibly out of reach.'

Alan Morrison, Poor Knights Islands, 21 August 1982: 'After about 40 minutes I was starting to tire and turned back towards the boat. Twice I felt a gentle tug on my fins. Upon turning I saw nothing. When I got back to the *Norseman* I was told that two of the dolphins (bottlenose) had been right behind me (witnessed by two people) and one had pulled my fins.'

From observation of peculiar markings, scars and fin patterns it has been possible to trace the movements of certain individual dolphins over periods of up to seven years (Simo, Busy Bee etc). At Whale Island a resident group of common dolphins was studied for a year by R. Stewart using the technique of photo identification of dorsal fins. We surmise that the toothmarks many dolphins bear are made by other dolphins as part of sexual foreplay akin to love bites.

Matt Conmee, Poor Knights Islands, 21 August 1982: 'At one stage I noticed one dolphin (bottlenose) mouthing another about 30 feet in front of me — an action like puppies biting one another's necks as they are running. It started up on top of the recipient, just behind the blowhole and ran its teeth down the right side onto the belly.'

Wade Doak, Poor Knights Islands, 5 May 1984: 'Twenty feet down, an adult bottlenose was lying draped over a flat, weed-covered rock. It was belly up, looking at me. I could see it was a male. Another dolphin had just left and a third came down and nuzzled its underside from tail to head while a fourth hovered alongside.'

Jacqui Samuels, Cavalli Islands, 24 June 1983: 'A couple were on the bottom, one lying on its back, and the other over its mate's stomach. Whether mating or just playing, we were not sure.'

Sometimes sexual foreplay may involve vigorous head-on charges, and might readily be mistaken for aggression.

Roger Grace, Moko Hinau Islands, 6 January 1986: 'Slightly separated from the rest of the group (20 bottlenose), John Young saw two apparently fighting. He summoned me over and there in the clear water about 40 feet down were two dolphins standing partly up on their tails and snapping at each other, sometimes actually locking jaws and making clear "barking" noises at each other. It went on for at least three minutes.'

Similar behaviours have been observed with spotted dolphins in the Bahamas. Copulation has been observed with both bottlenose and common dolphins.

Roger Rawling, Kaikoura, 12 January 1982: 'There were five dolphins (bottlenose) in the group. Two were swimming close while the other three kept their distance. All of a sudden the male rolled on his back and began swimming upside down. He then came under the female and for a moment they were locked together with their two front fins, their speed slowed down. After about 15 seconds they parted. This was repeated three times, the other dolphins keeping apart. The group then came together and swam off.'

At Kapiti Island, 6 March 1981, scuba diver Mike Young, observed common dolphins: 'At 60 feet, three pairs were swimming rapidly around me, belly to belly. They were mating. I could clearly see the male's penis thrusting and retracting as the pairs circled me no more than four to eight feet away. After a circuit they would separate, shoot back up towards Bruce (on the surface) and commence copulating while circling him.'

At such close range this explicit sexual activity would seem to have a communicative aspect. A similar episode with three pairs of common dolphins is illustrated on Plate 4A of this book (Mayor Island, Jill Gray).

Dolphins are notorious hybridisers, both in captivity and in the wild. Occasionally they have behaved as though extending courtship behaviour towards humans. This has advanced to a much more intimate stage by solitary male dolphins, as will be seen in Part Two, but it is sometimes manifested by group dolphins, too.

Near Red Mercury Island, 11 February 1986, Thelma Wilson and her companion, Dave Gibson, had had half an hour of strenuous snorkelling activity with a group of 30 bottlenose dolphins. There had been close approaches and exchanges of fancy manoeuvres, with individuals making intimate eye contact. Then, Thelma writes: 'One started circling me and had me a little concerned. He seemed larger than the others and was circling close and fast. Each time I dived down he barrelled over the top of me, then leapt out of the water coming down with tremendous force — enough to squash me if he miscued.' Thelma added that the possibility of rape *did* cross her mind.

While not relevant to this text, our files record feeding observations ranging over four species and show distinctive patterns for each. On three occasions, common and dusky dolphins have given evidence of stunning their prey, as per the sonar hypothesis of Dr Kenneth Norris.* While common and dusky dolphins have been seen intermingling, it appears there may be mutual avoidance between common and bottlenose dolphins. The former have been seen with baleen whales and dusky dolphins, and the latter accompanying spotted dolphins, orca and pilot whales.

Several interesting reports involve dolphins observing human activities underwater. As a wreck diver on a strict time schedule has put it — he could not leave off his work and *approach* the dolphin, almost a reflex action for most people and not one which dolphins appear to accept first off. In this way, busy humans may be more approachable for dolphins and intrinsically more interesting than passive humans.

We have accounts of dolphins, alone or in groups, watching divers as they excavate a treasure ship, lay sewage outfall pipes or set up seafloor experimental apparatus; dolphins watching above water as men repair an upturned boat or deliberately crash man-powered flying machines off a jetty into the sea.

Observation of human activity is best in situations where the performers are less likely to interrupt their tasks on seeing dolphins in the vicinity.

Since this report was prepared, Project Interlock has collected many more anecdotes of human/dolphin encounters. For the most part the broad pattern

---

*Dr Norris has stated that there is circumstantial evidence for his theory that dolphins and some whale species may stun or kill their prey with high energy bursts of sound. He marvels at the ethics involved, should it prove true that each of these creatures is possessed of a lethal weapon never used on its own kind, nor its major assailant — man.

that emerges here continues to unfold, but there is a tendency for the quality of meetings to improve, perhaps as a result of feedback. We hope that this book may serve as further stimulus.

PART TWO
Another Window on Ocean Mind

In just my lifetime, the world has changed drastically. From a place where people were surrounded by wild animals, the world has become a place where wild animals are surrounded by people.
William Conway

For a decade Project Interlock has assembled a file of cases where dolphins, apart from their social group, have lived in close proximity to humans. In such situations complex interactions may develop involving interspecies gamesplay, acoustic exchanges and body contact. These activities have often led to play-bonding of considerable physical and emotional subtlety between humans and cetaceans. While not exhaustive, we now have 30 written accounts of such episodes. The majority (26) have been with solitary dolphins but the phenomenon cannot be so simply defined — two cases have involved pairs, and two have been with mother/baby sub-groups.

Although 20 have been with bottlenose, our file extends to three other species: spotted (1), common (2), and dusky (1). They have been both juvenile and adult; male (12) and female (9). In eight accounts, the gender has been uncertain. The most complex and prolonged human/dolphin associations have been with bottlenose (also the most coastal in habitat), but similarities in the responses of common, spotted and dusky dolphins would indicate the phenomenon is not limited to a single species.

Why gregarious animals like dolphins should live apart from their social groups is still a mystery to us. Perhaps the company of humans offers some compensation for solitude. With some famous cases, such as Opo in New Zealand and Donald in the United Kingdom, the dolphins have been described as 'social outcasts' or 'bulls ostracised from the herd'. Before sufficient data has been assembled, such simplistic explanations do not seem justified. Perhaps it is better not to regard the phenomenon as some sort of aberrant behaviour, but rather to study it in much greater depth with an open mind. Opportunities for prolonged free association between our species and dolphins are all too rare. Much can be learned about dolphin capacities in such flexible situations. To the participant observer lone dolphins provide another window into the cetacean mind, undistorted by the insult of capture and detention.

I have found it difficult to discuss this subject without creating a term that covers all aspects of our 30 episodes. Therefore I wish to use the partial acronym, DINT, to refer to solitary or sub-group dolphin-initiated interaction with humans; or it might be called *The Classic Opo Phenomenon*.

History of DINTs

109 AD (approx)	Hippo & partner	Tunisia	Species unknown	
1814	Gabriel	England	*Tursiops truncatus*	male
1888-1912	Pelorus Jack	New Zealand	probably *Risso's*	
1953	Fish & Hoek	Sth Africa	*T.aduncus*	female
1954-1955	Opo	New Zealand	*T.truncatus*	female
1960-1967	Charlie	UK	*T.truncatus*	female
1961-1962	Wallis	Australia	*T.truncatus*	uncertain
1964 Currently in progress (1st report 1978; 12 in all)	Monkey Mia pod	W.Australia	*T.aduncus*	males & females
1965	Nudgy	Florida	*T.truncatus*	male
1972 (approx)	Nina	Spain	*T.truncatus*	female
1972-1978	Donald	UK	*T.truncatus*	male

1976-1978	Sandy	Bahamas	*Stenella plagiodon*	male
1978-1979	Horace	New Zealand	*T. truncatus*	male
1978 Currently in progress	Jean Louis	France	*T. truncatus*	female
1978	Elsa	New Zealand	*Delphinus delphis*	female
1979	Dobbie	Red Sea	*T. truncatus*	male
1981 Currently in progress	Whitianga mother/baby	New Zealand	*D. delphis*	female & baby
1982-1984	Percy	UK	*T. truncatus*	male
1982	Indah	Australia	*T. truncatus*	male
1983	Costa Rican	Costa Rica	*T. truncatus*	male
1984	Tammy	New Zealand	*Lagenorhynchus obscurus*	male
1984-1985	Simo II	UK	*T. truncatus*	juv. male
1984 Currently in progress	Rampal	New Zealand	*D. delphis*	male
1984 Currently in progress	Dorad	Ireland	*T. truncatus*	male
1985-1987	Romeo	Italy	*T. truncatus*	male
1986 Currently in progress	Fanny	France	*T. truncatus*	female

Footnote: In 1988 we learned of four more episodes: Billy, Australia; Herbie, Bahamas; and others in Spain and Yugoslavia, for which we await details. ('Currently in progress' refers to early 1988.)

First Glimpses: Elsa

My interest in solitary dolphin behaviour was rapidly aroused when two episodes occurred in New Zealand within the same time frame — one on my doorstep at Ngunguru, the other at Napier.

Three days before Christmas 1978, a black fin was being followed gently by two boys in a dinghy out on the Ngunguru River. Among a line of boats and moorings, my wife Jan dived, and dolphin and woman met at close range. We were staggered to find a common dolphin, creature of the open sea, cruising in an estuary.

On the river edge a group of children were calling, 'Here dolphin, here dolphin,' their hands outstretched. The dolphin moved towards the children. We could scarcely believe our eyes when it swam through the shallows right up to the children and grounded. It didn't appear to be in the least afraid. Children were milling about straining to pat and stroke the sleek form. Sang one little girl with shining eyes, 'I've never touched a dolphin before.'

The total trust and passive self-surrender were quite uncanny. Even the beach dogs were barkless and spellbound as the children lavished their affection on the lustrous creature. I was so interested in watching this spontaneous child/dolphin interaction that I didn't involve myself for some time.

Jan was warning the kids not to put their hands near its blowhole. Gently they splashed water over the glistening skin and she told them to avoid getting it near the blowhole when it was breathing. Faces alight with pleasure, the children were wide open to suggestion and ready to learn. I began to whistle to it. The dolphin replied, with inexpert wheezings through the blowhole. Obviously a young dolphin, it was attempting to respond to my whistles with air-borne sounds.

I would have liked to explore this avenue at greater length but there were others to share the visitor with and it revelled in the attention, showing no concern at being stranded with the falling tide — seemingly confident that we would give aid.

Worried that it might get hurt as the tide receded I eased it out into deeper water. As I turned to walk ashore the dolphin rolled on its back playfully displaying its white belly. Then it came ashore again a little further upstream. This time we examined it carefully for any problems. The only blemish was a faint half-inch scratch on its back just in front of the dorsal — a perfect dorsal, all black with a tiny nick in the trailing edge near the tip. We decided we could take the liberty of rolling it over to see what sex it was — a young female, the anal-genital opening flanked by two mammary slits, like a division symbol $\div$.

To determine her length I stretched out alongside her on the sand. We matched — five feet six inches. She seemed to respond to this manoeuvre with pleasure, perhaps a nicer way to encounter man than meeting a forest of legs.

The tide was falling faster and we decided she should return to her element. There was a considerable distance of shallow sandy plain with only two feet of water covering it. As I guided her out I felt a wave of anxiety — a sympathetic response to her predicament. While submerged I was unable to determine which way to head to deeper water and the tide was receding every moment. But I raised my head above water and used my knowledge of the landmarks — the physical features of Ngunguru estuary — to guide me. Did the dolphin have this knowledge?

When we reached deep water, I released her and we swam along together for a time. Out in the racing mainstream our friend Terry met her with his underwater camera. Despite the sandstorm, he managed to capture a close-up portrait and then passed his camera out to gambol with her.

After several exchanges of body language the dolphin came wafting down-current towards him. He lay on the surface, both arms extended, palms up, fingers outstretched. Then to his astonishment: 'She *touched* me, turned my hand over with a gentle muzzle, rested on my wristwatch and slowly rocked her head laterally five or six times — tickaticka tickatick — ten seconds touching.'

As the tide receded further she began to hunt at a point where the river courses through a rocky race, a spot where mullet would be easy targets for an acoustic creature. I decided to stay with her until I knew the outcome. Would she return to sea or advance back up into the estuary?

When I left at 5pm she had disappeared beyond the river entrance. I felt sure she had returned to sea with the tide.

I learned later that this was not so. She must have returned with the incoming night tide, and was sighted cruising among the moored boats in the dark. In the early hours of the morning, it seems she was stranded among sharp, oyster-covered rocks near the river mouth. We found Elsa on the river bank next day, her left flank deeply lacerated with oval-shaped gouges right through the skin to muscle level. Sea lice had attacked the wounds. Beneath her chest linear scratches showed where she had struggled over a sharp surface.

When Fisheries biologist, Lew Ritchie, came by and saw her predicament he hurried off to get his boat and a veterinarian so we could take the dolphin to sea and release it.

Meanwhile Jan, Terry and I took turns to hold her head clear of the rocks, assisting her to breathe and comforting her with our voices. Clearly she was in a state of shock, but she gradually calmed down. Jan, who is a nurse, could feel the heart-beat and knew when she became agitated. As people came wading up to look, her heart would speed up and she would start wagging her head back and forth in a searching manner as though trying to locate the direction the thing was approaching from. As people left and everything became silent, she calmed right down.

When Elsa Smith arrived, a strange thing happened. Elsa was bending down with her hand hanging in the water beside the dolphin's head when it deliberately rubbed her hand several times. And so, the name Elsa.

The sea was rough offshore and it was three hours before Lew arrived with his husky dive boat. The moment I heard the roar of his stern drive racing over the river bar, I thought, 'Thank goodness, he's made it.'

At that instant there was an explosion, a loud bang from Elsa. I was just in time to see her do it again. She had retracted the sphinctre of her blowhole, revealing delicate membranes and muscles deep within. A fold in the skin was inflated like a cherry — it bulged. Then with a deliberate crack she burst it.

When the vessel arrived, we steered Elsa out to the knee-depth hull.

Alongside, she made another volley of cracking sounds with her blowhole. Terry got the hydrophone and tape-recorder going. Just before we lifted her aboard, a series of high-pitched underwater vocalisations were recorded. Then, without the least struggle, totally relaxed, she allowed us to lift her aboard on to a foam rubber squab. We covered her with wet sacks, soothing her with our voices, as the vessel carried us over the river bar and headed straight out to sea. When the waves became violent enough to jolt the hull sharply, we thought it best to stop and release her in case she suffered internal damage from the boat's action. The vet gave her a shot of antibiotics in her back near the dorsal fin. She didn't flinch. We waited for any shock reaction before lifting her over the side. She let out a series of deep sobs from her blowhole, like a child trying to be brave, and nodded her head. Then she was calm. Her suffering and self-control stirred us deeply.

We eased her over the side into moderate following seas. She swam on the surface with no difficulty. We hovered by as she made a series of circles, as though orientating herself, then she set out into the seas, on a northerly course.

About a mile ahead Terry glimpsed a flock of gannets wheeling. There was a chance she might locate her companions out there beneath the diving birds. We all hoped she would meet up with them because she was so vulnerable to shark attack by herself. The seas were rough, poor Terry became ill and we were drenched with spray. Lew headed for the nearest haven — Tutukaka.

Horace

A few months before Elsa swam into our lives, a lone dolphin had arrived down in Hawke Bay. It set up station around a marker buoy off Westshore beach, not far from New Zealand's only dolphinarium. With spectacular tail-standing leaps it soon attracted the attention of the people of Napier. It was a male bottlenose about ten feet in length. Frank Robson named it 'Horace' after Horace Dobbs, whose book about another lone dolphin, Donald, had recently been published.

By early December 1978, the dolphin's antics had reached the notice of the local press: 'Horace Steals the Show'. A small circle of enthusiasts began to devote their lives to observing its movements, providing it with the companionship it seemed to seek. Bottlenose dolphins are not common along this stretch of coast.

Once it had achieved a rapport with the locals, the special leaps and friskings ceased. Among the devotees was Quentin Bennett, one of our oldest diving friends. A superb water-person, Quentin had shared dolphin experiences with Jan and me aboard our catamaran as recently as the previous month. Now he began sending us 'Horace' reports.

Every time Quentin went out into the bay he met the dolphin in the same small area. Usually, because of his work, it was in the early morning or late evening. Until they had been playing together for a while, Quentin found the dolphin suspicious of hands and arms so he avoided reaching out for him, leaving any contact to Horace.

Entering the water clad in a wetsuit, he would dive to the bottom clicking a metal noisemaker such as you find in Christmas stockings, and making motorbike noises with his mouth. When Horace arrived, he would give him a chance to scrutinise him thoroughly with his sonar while submerged. Then he would indulge in the wildest antics — only a flexible, fit diver could match Quentin in this respect. Chasing, turning upside down, cork-screwing and haring around; jack-knifing under, rising rapidly out of the water, dolphin and diver strove to outdo each other in boisterous mimicry. Quentin would spend as much time as possible underwater and the dolphin tried to impede his return to the inflatable. He wished to play actively the whole time and quickly got bored.

More and more friendly trickery entered their gamesplay. One afternoon Quentin took Perry Davy, a young boy, out to meet the dolphin. As they became chilled Quentin suggested they return to the boat. Horace approached Perry and led him gradually on to a changed course until he finally had the boy swimming in exactly the wrong direction.

As summer advanced, Horace became fond of escorting any yachts and fishing boats leaving Napier, north as far as Portland Island and south to Cape Kidnappers. He frolicked with any small craft that came near him. By 19 January, he had begun to accompany yachts up the river into the inner boat harbour. As a yacht was entering the channel, Horace nudged the rudder and altered its course back to sea. Other yachtsmen found their rudders strangely immobilised. He would wait for small boats at the jetty and as they were preparing to tie up, he would swim over and shove with his beak to push them off course.

In Napier's main harbour Horace was a favourite with the construction divers and fishermen. He would steal the divers' flippers and flick floating onions around with his beak. He accepted live fish tossed to him from fishing boats and would take a fish from a trawl net, playing with it before swallowing.

In the course of our research we were delighted to receive a letter and photo from Pat Wellgreen telling of a meeting with Horace at sea. Although it belongs further on in the chronology, I will present it here.

'Our dolphin experience happened on 20 May. Horace was known to follow people out to sea, and so we were overjoyed when he chose to accompany our 12-metre trimaran. While we were motoring out of the channel he kept rubbing himself against the rudder. As we got further out into Hawke Bay, he became more used to us, and came up alongside. Someone suggested that we use the blunt end of the boat hook to splash in the water to attract his attention. Horace thought this was "okay" and he had a rub against it. We played with him and the boat hook for ages. He then decided to do a few flips for us. It was fantastic. Much better than Marineland! He seemed to take a perverse delight in splashing us all.

'We had hoped to do some fishing, but Horace's fishing was much better. He chased a large kahawai round the boat, between the outriggers, as if he was a cat after a mouse. Eventually he caught it and dragged it backwards by holding its tail in his mouth. When it was quite dead, he swallowed it head first. All the while we were running round and round the boat yelling and clapping.

'Horace stayed with us most of the day. However, just as we were heading for home, he saw a fizz boat and went over to say "hi" to them.

'That day, we were told, quite a crowd had assembled at the sailing club, ready for a demonstration by Horace. It was a shame that so many people had missed out, but we had such fun with him that we couldn't feel too sorry.'

Gradually Quentin began to notice a peculiarity in Horace's play — mirror-image mimicry — which suggests a subtle level of abstract thought. The pattern first began to emerge on 26 December, when Quentin tried to teach the dolphin to leap over his inflatable, a game Donald used to play in Britain. Swimming rapidly up from the bottom to the side of his inflatable Quentin slithered aboard, strode across and dived headlong in again. But to his dismay, Horace did not follow. In a dinghy tethered behind, Quentin's small daughters, Annika and Camilla, were convulsed with laughter at the dolphin's antics, as he totally fooled their father, who wrote:

'Horace was right behind me so I couldn't see what was happening. As I went over the boat he turned *upside down* and slid under it, in effect, mirroring my whole act. Because of the usual Napier murk I hadn't the least idea as to what was going on but the children could see it from the dinghy.

'Every week we are hearing more stories about Horace's games and antics. One thing that appears increasingly is this mirroring of what people do. He plays mostly with yachtsmen and quite often gives unknowing sailors serious frights by banging on the bottom of their boats when they don't expect him. Of course he is always pushing up centreboards and interfering with rudders.

Apparently if he pushes up a centreboard and the sailor slams it down, Horace slams it back up. If it is pushed down slowly, he thrusts it back up slowly . . .

'I had a friend out following the National Flying Dutchman Champs from a little put-put boat. Horace started playing around so he steered his boat on a weaving course. Horace adopted a weaving course too — mirror image:

'The other day he alarmed a young girl in a yacht who was not very experienced and rather worried at his antics with the centreboard. He came up beside her and she threw water on his face — "go away". He turned face down and flicked water over her with his flukes.

'Apparently he has tipped a small yacht over in the inner harbour. It must have pricked his conscience a little as he came back to the unfortunate sailor and checked that he was all right, and the chap was able to touch him. The Governor-General, Sir Paul Reeves, when Bishop of Auckland, has had the centreboard of his Sunburst dinghy split by an over-zealous Horace.'

By the end of January, Quentin was often touching Horace briefly. The dolphin would spend part of his time playing around in the main harbour where the working divers enjoyed his company. On 7 March underwater blasting operations began, to deepen the port. Each time the chief diver telephoned a warning, former dolphin-trainer, Frank Robson, would entice Horace out of the danger area with his boat, into the safety of the river and the inner harbour.

On 8 March, Quentin set out to get the first underwater pictures of Horace. Up till then, although an ardent photographer, Quentin had not attempted to use a camera at any meeting. They found Horace near a trawler fooling with a large flatfish. After shooting two rolls of highspeed film with an ultra wide-angle lens, Quentin decided that that was enough — time for their usual play session.

'I played the games of my life. We put our noses together. Horace's beak against my face mask for ten seconds at a time. He towed me around in various ways — me holding his dorsal fin; me holding his chin; me holding his pectoral fin; he holding my hand in his teeth. We played more complicated manoeuvres — he sliding his body along my fins, me doing similar things to him. The usual barrel-rolls and somersaults — body vertical in the water, head up, head down; face to face — shake heads; and he screaming around me making the motorbike noises that I do to him at times.'

'It was something the like of which I've never heard before. Something that makes me feel honoured to have been part of. I was careful not to hurt him — it worries me holding his fins.

'Quite often when playing one sees that he has an erection.* At these times he is quite aggressive in a sporting, challenging manner. He will try to lead you from the boat and trick you.

'Yesterday, after I put the camera away, it was something totally different, almost like a ballet; mutual feeling and respect were certainly part of the episode. No erection.'

Quentin is a busy person and not a wordy letter writer so I was amazed to hear from him just three days later. Dolphin lover, Graeme Thomson from Wairoa, had arrived wanting to meet Horace. They searched all his usual

*Quentin notes: 'Erections were common in our early days but were not evident later. Perhaps he had found out that I was straight.'

66

spots out in the bay and then headed back to the inner harbour — where he had been all the time. Hoping for clearer water, they led him out but it was still very murky with less than one foot visibility, which makes it difficult to play underwater with a dolphin.

He would come up underneath the divers, gently pulling first one flipper, then the other, turning the diver in a circle, and manoeuvring his legs apart.

'I had a while with him,' wrote Quentin, 'warming him up. He is careful and needs quite a bit of play before he gets involved. He and I got quite involved. I held him several times and finally he came under me and lifted me out of the water astride him and gave me a ride! I was in front of his dorsal, arms raised above my head in sheer exhilaration. We were all incredulous. Then he gave me a tow, holding his dorsal fin. The murk made it difficult to see what we were doing and we were afraid we might accidentally kick him in the eye with a fin, and put him off humans for a while. I was cold and had to get out.'

Quentin was just about to leave on a journey to Sweden when late in the afternoon (28 April) he, with his wife Tina and some friends, made a last visit to Horace. By this stage Horace had established court in the inner harbour around the yacht *Tiny Dancer*, where he met a regular set of human friends.

'We all had a great swim with Horace, about half a dozen people with him at the same time. I had, perhaps, a dozen or more tows, mostly along the bottom. Visibility was about 12 inches and there were three boats around the whole time. At one stage I jack-knife dived. Horace and I collided beak to mask. It shook me and I was quite stunned — almost knocked out. Except for this collision, which I am certain was a mistake, he never once hit me, although he did hit others, rather shaking them up. This intrigues me, as I always played rough and energetic games with him. My type of water play really gets Horace going at high speed, buzzing in head-on and jumping out of the water. We have familiar routines and the murk doesn't matter so much. Other people in the water get frightened at this and think he is aggressive. I feel he is just boisterous and I give it back to him — unless it has been me that has been the boisterous one and he is "giving it back to me".

'Then I got a surprise. Tina has never swum with Horace before. She hopped in and he came up from behind in the nil visibility water. As she was still facing the boat he could not have approached from the front. Gently he nudged her in the small of the back. Instantly she turned as no doubt he expected, and he took her right hand in his mouth. Now, Tina is rather chary of animals (she is a superb diver) and no animal, not even a pet belonging to her closest friend, would be permitted to take her hand in its mouth. Yet she happily let this so called "wild animal" do it without thinking, on their first meeting, in filthy water. Why did Horace wish her to turn and why take her hand? Did he understand the human significance of this manoeuvre? It was something simple and meaningful to any human being from child to academic, and something special.

'At every meeting with Horace, humour seems to have a part. I wonder what intellectual level this capacity would indicate to a psychologist?

'There is a young couple, Rosamond and Allan Rowe, who go down to the inner harbour with their two children almost every afternoon after work to meet Horace. Ros — a lovely, natural woman with an incredible feeling and affection for Horace — keeps records of each interlock. He has given her three rides recently.'

Rosamond Rowe: On Quentin's suggestion I wrote to Ros and Allan Rowe, and an extensive correspondence developed. Eventually I flew to Napier to meet them. Here is a letter I received from Ros:

'I hardly know where to begin regarding my friendship with Horace. Our relationship — meaning my husband Allan, our two children, Selwyn and Odette, and our elderly friend, Miss Bingham — has been on a simple level and is a rather personal thing but we have in the course of our interspecies "love affair" made some rather startling observations of his responses to us.

'As regards my experience as a swimmer or diver — well, I can swim a few yards if necessary and there my qualification ends. As the sufferer of a middle ear disease, I am not really supposed to go in the water, but I feel safe and happy bobbing around on the surface in my wetsuit. It seems to be no handicap to my enjoyment of Horace or his acceptance of me.

'As regards previous experience with animals, I have always been surrounded by them, both wild and domestic, and sometimes wish I could take off to the hills to commune with all other living things.

'When we heard through the news media of the arrival of Horace, our family spent all the Christmas holidays trying to meet up with him. It was not until around 18 January that we first met him on the sea's edge off Westshore beach, and played in the breakers with him.

'Early in our friendship (18 February) my husband Allan swam with him several times out in deep water in the bay and the two enjoyed games of chase and hide-and-seek with seaweed. Horace would come up to him with a piece of bubble kelp in his beak. When Allan grasped one end, Horace tugged until it snapped. He then dived to some depth and released it, returning to challenge Allan for the other half. This continued until only a shred remained. By this time the first half had floated up and the game resumed.

'Horace must have learned to associate Allan's voice with these games and he started coming up the channel into the inner harbour whenever Allan called him from the shore. He would dart off and return with a piece of seaweed.'

For five months the Rowe family devoted their lives to following the movements of Horace and from 20 March to 26 May, Ros kept a log of all sightings and interlocks with him.

Over the 68-day period, they recorded 39 sightings of the dolphin of which two-thirds were in the vicinity of *Tiny Dancer* in the inner harbour. The rest were in the river channel (6); the bay (4); and the main harbour (5). Human/dolphin in-water exchanges occurred on 22 occasions. They noticed that when Horace was actively feeding in the river he did not want company and would indicate this with rapid tail slaps if approached.

On average, Horace stayed up to three days at a time and the average of absent days was about the same with the longest stay seven days and the longest absence ten. Prior to one lengthy absence there had been a severe gale.

While there were other people interacting with the dolphin, Quentin Bennett and the Rowe family appear to be the most intensive from an in-water, interlock viewpoint. Ros Rowe's letters highlight their most exciting encounters:

Feet Upon a Rock* — 'One incident which is clear in my mind, and always '

*Title of an autobiography by Rosamond Rowe (Caveman Press).

will be, took place on Wednesday 28 March. This was before we had purchased our wetsuits and rowboat and I was forced to stay within my depth because of my silly, giddy head. Horace was playing with the rudder of his favourite yacht, *Tiny Dancer*, moored 30 metres from the Sailing Club beach in the inner harbour. The temptation to get out to him was great so I threw caution to the wind and got an acquaintance to tow me out to the yacht with his catamaran.

'Once there, I clung precariously to *Tiny Dancer*'s rudder and watched enchanted as Horace swam around close to me. At this stage I knew him as a fin and, to him, I was only a voice which had called his name wistfully from the shore. After a while the effort of clinging to the slippery rudder became exhausting, so I called Allan to swim out and help me ashore. Once he reached my side I felt able to relax my grip a little, so I stretched my stiff and weary legs down to prepare for the trip to shore.

'To my surprise and utter relief, I felt a solid rock beneath my feet. I was able to have a breather. I said to Allan, "What a clot I am. All this time I've been hanging on for dear life and there's been a rock to stand on."

"No there isn't, you dope," was his amused reply.

"Well, what do you think I'm standing on then?" I retorted. I stretched my legs down again to prove my point and was astonished to find there was nothing solid there. I had been standing on Horace.

'After a spell on shore I saw that he had followed us into the shallows so I raced in again. This time as he glided past, I stretched out my hand and touched him for the first time. The feeling of awe that Horace had supported me when in need, and stayed stock still when to have slipped from under me would possibly have spelt trouble, giddy as I was, deepened next morning. Writing it up in my diary I felt moved to abandon the task for a moment and pick up my well-thumbed Bible. I opened it at random and my eye fell on a verse from Psalm 40: "He set me safely on a rock and made me secure".'

Stroking — 'The 14 April encounter was special because it was the first time Horace presented himself to me to be petted from a boat. I felt my previous physical handling of him was with his acquiescence whereas this time he was actually *inviting* me to touch him.

'We went to *Tiny Dancer*'s mooring in the inner harbour to see if our friend was still present. And there he was bumping the yacht's rudder, so we eagerly launched our little boat and the four Rowes hopped aboard to go and ask Horace if he wanted our company. As soon as we started rowing he left his yacht and came alongside us. I rolled up my sleeve and cautiously extended my hand to see if he would resume this morning's game. (Ros had clicked her fingers in the water and Horace had repeatedly nuzzled her hand, both gently and boisterously.) He inspected my hand and let his body slide past it. Then he lay passively just below the surface allowing me to stroke every part of his body, at times raising his head right out while I cupped his beak in my hand or stroked the top of his head. It was very moving for me and I shed tears all over him. He would occasionally swim away from the boat and return upside down, inviting me to rub his underside. Once as I was turning my head to speak to the kids, he came up smartly and nipped one finger hard. It hurt quite a bit and I told him I didn't think much of that trick. Next he came back with his mouth open seeming to dare me to trust him by putting my hand in his mouth — so I did. I received several more nips but none as sharp as the first.

'My parents were on the shore so we brought them out to watch as Horace presented himself for more caresses. It is funny how, after a particularly rewarding communication with Horace, I have dolphins leaping through my dreams all night long.'

First Ride — 'The first time I was given a ride by Horace was well timed in a very special way. Earlier that day (23 April) I had heard the distressing news that the beloved companion of my childhood, my horse Giselle, had died. Feeling rather upset, we went down at four-thirty as usual, to look for our friend. There he was in a very trusting mood. He allowed both Allan and Miss Bingham to touch him. I decided to experiment with singing to him as I had been wanting to try it for some time but felt a little foolish in public. I sang "You are my sunshine", and Horace seemed enthralled (which says little for his appreciation of the arts) and he lolled placidly beside the boat, his pinhole of an ear above the surface to listen.

'Then I climbed out of our little rowboat into the water with him. He immediately swam beneath me so, on Allan's suggestion, I spread my legs as if straddling the back of a horse. Horace swam beneath me, then rose up gently until I was astride him. I was not sure where I was sitting until Horace came up to blow and then I discovered I was between his blowhole and fin. I was borne carefully and majestically through the water, completely overwhelmed by the experience. The ride ended when he gently swam free. He played around a bit more before repeating the procedure several times. A surf ski whizzed past and Horace shot off like a rocket to pace the man on the ski.

'To our amazement Horace left the ski by the Iron Pot and headed back to us. I leapt out of the boat again and was joined by a different Horace. This time the party got a bit rough. I was nipped on the leg, butted in the behind and herded in tight circles. The water was churning and boiling, and I was just flotsam completely at the mercy of our huge friend. At one point, he leapt out of the water and entered again, his large tail within inches of me. A boy arrived in a dinghy and Horace darted over repeatedly, slapping his tail on the surface. Strangely enough he did not resent Allan's presence in our boat. I looked enquiringly at Allan knowing he would tell me to get out if he thought things were getting out of hand. Then I decided to wait and see what developed. Horace remained excited and I was starting to tire, so I landed.

'A short while later Horace turned up again just behind the boat. I got in again, as much as anything to try and end the interlude on a quiet note. Fortunately Horace seemed to feel the same and he calmly greeted me before slipping away to inspect *Tiny Dancer*'s rudder. I swam ashore.'

Miss Bingham Possessed — 'Our friend Miss Bingham, who is 72, enjoys meeting Horace and we had always thought the dolphin respected her advanced years, treating her with a special gentleness. However, on 8 May, he exploded this myth in no uncertain fashion.

'One evening he turned up in the inner harbour after an absence of ten days. Miss Bingham excitedly jumped in to swim with him and by the time I had done the same, she was being tossed around pretty roughly by Horace. He was sexually aroused and not responsive to her attempts to settle him down. She reluctantly left the water. I decided to stay and try to quieten him and was successful for only a short time before I too was being buffeted around.'

Note: Miss Bingham sent us her own account in which she suggests that her helplessness in wearing a buoyant wetsuit with no leadweights may have been a contributing factor in the episode, coupled with the long absence of their friend. Her account is as follows:

'My excitement was intense and Horace must have picked it up, as the prods with his beak became more and more insistent. He was determined to make this ridiculous creature play — the prods and butts became harder and the lifts more spectacular. Inspite of laughing and enjoying it I knew I had to get out. Allan was laughing so hard and trying to take a photo he did not realise my calm request for help was really a shriek of near-desperation. After one last terrible jab in the ribs I was able to grab the boat and scramble on to the rocks with, so they told me, Horace following hopefully.

'Ros swore it was a clear case of attempted rape and Allan said, "Well, you did have your wetsuit flap trailing provocatively behind you." In my haste I had forgotten to buckle it up before getting into the water. In any case, my ribs were sore for months afterwards. Some days later I met him again, with my leadweights on, and he behaved like a perfect gentleman.'

In her letter Ros Rowe concludes the incident with the comment, 'Horace has been contrite ever since. In fact, in all the times we have played with him he has acted that way only twice — the other occasion was on the evening of my first ride. Possibly my excitement in having this experience was communicated to him.

'On 21 May, we enjoyed, from our boat, a lovely interlude with our friend. We met him in the part of the inner harbour called the Iron Pot. He was in an extremely placid mood and showed every sign of delight as we rowed into the area. He quickly swam over to us and there he stayed a very long time. He lay beside our boat and lapped up our caresses and words of endearment. Selwyn, our eight-year-old son, loves to stroke Horace but sometimes misjudges his distance and places his hand a little close to the sensitive areas of the eye and blowhole. Horace never seems to mind, and on this occasion appeared to want Selwyn's attentions, squeezing his eye shut in an exaggerated fashion and staying beside our son instead of moving away as one would have expected.'

The following night the Rowe family had their last intensive interlock with Horace. He gave Ros numerous fin tows, coming up from below, nudging her gently and nipping her hand affectionately. When she got out, he seemed to want her to return and gurgled like a human baby when she did. Then he took her for the longest tow ever — so far out she let go and swam back.

When Miss Bingham got in, he would approach her gently from behind, put his beak into the small of her back and push her quietly across the surface.

There followed only two more in-water sessions, on 23 and 26 May. On the final encounter, in retrospect, Horace seemed nervy and different. There was little physical contact apart from nudging Ros's hand several times. And from that date nobody saw Horace again.

Miss Bingham thinks she saw him at dusk on 7 June where she lives, a few hundred metres from the main harbour, feeding about 30 metres from the coast. That night an enormous blast shook the houses in the vicinity of the wharves. This time nobody had given a warning to entice Horace to safety.

Three days later there was a report from the yacht *Trinity* of a huge fish floating on the water a mile out beyond the pier heads.

But Horace's fate is not known. He may yet turn up among the bottlenose dolphins to the north where his species is more common. Or could it have been the diesel spill in the boat harbour on his 28 April visit? Oil on the lungs leads to slow death . . .

In the history of lone dolphins approaching human settlement, one thing is clear — it is highly dangerous for the dolphin.

A History of Dolphin Initiatives (DINTS)

The Elsa and Horace episodes prompted a new branch of our research: parallel with our studies of dolphins in social groups, much might be learnt from solitary dolphins once they establish a lengthy relationship with humans. As mutual trust develops valuable insights into their capacities could emerge.

As a New Zealander I was already familiar with the saga of Pelorous Jack; and the story of Opo is a golden summer memory to people of my generation — a fifties pop song on the radio and tales filtering through from our far north, of beach frolics with kids by a playful dolphin.

This was the context for setting out on a lengthy study in which I found more and more such episodes. Now I wish to go back to the earliest written accounts and trace the development of Dints through time and space. This is a global phenomenon that has undoubtedly existed in many societies where there have been no written records.

In this review I hope the reader will look for recurrent patterns and consider the implications of so many flexible and creative behaviours as we try to grasp the nature of non-human intelligence.

Clearly, humans share with dolphins a mutual curiosity and some degree of emotional empathy, as they do with domestic pets. It would seem that solitary dolphins have been adopting humans as substitute companions, fulfilling in some measure their needs for the stimulation of play. For us, open-ended relationships with other species are all too rare as humans usually adopt a dominant role towards other life forms. As a channel for communication, the trustful relationship of play offers us unique glimpses into alien mental processes strangely akin to our own. It is not easy. During our adult lives we have been losing the awareness and sensitivity necessary to establish the trust of play. It is a great irony that this may be the limiting valve which inhibits some scientists from ever taking that first child-like step towards interspecies communication, or of accepting that it *is* possible.

As my review leaps from antiquity to modern times, an increase will be noticed in the quality and frequency of exchanges between solitary dolphins and aquatic humans; technology has enabled us to adapt our bodies to the ocean with insulating layers, improved propulsion and vision, and an extended air supply. Furthermore, the essential benign context for human/dolphin encounters that existed in antiquity has gradually been restored. Whether regarded as children of the gods or flagbearers of the environmental movement, whether pantheist or ecologist, positive belief systems provide a climate for interlock, after intervening centuries of cultural apathy when dolphins were at risk whenever in the vicinity of humans.

But what of the future? For the past 30 years scientists have been tuning radio telescopes to the stars in search of alien intelligence. As Doctor Gerrit Veerschuur of the National Radio Observatory, Arecibo, has said, after viewing film of a remarkable diver/dolphin exchange*: 'We will first have to mimic

*The Maravilla Dolphins: Bahamas chapter

any signal received from an extraterrestrial intelligence. The dolphin connection is worth further exploration for the insights we may gain from experiment with this form of interspecies communication. The cetacean species has demonstrated a willingness to communicate, and has highlighted an amazing inability on our part to deal with this.'

The history of Dints may provide clues for a major cultural leap from the oceans of our own blue planet to minds elsewhere in Universe. Scientists are agreed that the probability of our being alone is as absurd as the 'flat earth' theory. But would we know how to reply?

In his book *The Dolphin: Cousin to Man* (Penguin 1968), Robert Stenuit gives an excellent review of human/dolphin encounters in antiquity. The most interesting and detailed is that of a dolphin I have called Hippo which offers parallels with several Dints in modern times.

In the Roman period, around AD 109, Pliny the younger wrote of the Tunisian sea town of Hippo where a dolphin befriended a boy swimming offshore, bearing him back to the beach. The following day the dolphin returned, accompanied by another. But the boy, along with his companions, fled from the sea. For some days the dolphin leapt high above the water, frisking about and twisting its body into an 'S'. Gradually the boys lost their fear, approached the dolphin, called to it and played with it. The other dolphin remained in the vicinity, but only as a spectator.

A deep friendship developed between the boy who established first contact and the dolphin. Eventually its fame attracted crowds of visitors to Hippo and it was officially honoured. But resentment at the disruption of city life developed and the dolphin was secretly put to death.

From around AD 109 until the nineteenth century, very little in the way of dolphin/human encounter entered the Western written record. *This does not mean none occurred.* In the Pacific, just as in ancient Greece and Rome, dolphins were regarded as messengers of the gods. Ancestral spirits were believed to take the form of guardian dolphins or sharks, which guided canoes, affected rescues and assisted in fishing.

Around the world many cultures believe that any harm to a dolphin may cause ill-fortune or sickness. In the Ganges, Amazon and other major rivers inhabited by freshwater dolphins, similar protective myths have arisen, often coupled with symbiotic fishing practices. Mutual assistance in fishing is another widespread human/dolphin relationship that has now been well documented in both North Africa and Brazil. There are now so many records of dolphins behaving altruistically towards humans that stories of dolphins rescuing or assisting people can no longer be dismissed.

At some point in history the West rejected stories of human/dolphin involvement as the stuff of myths and legends. From an ancient civilisation, centred on harmony with natural forces symbolised by gods, there was a shift to a creed based on human supremacy, dominion over nature, and exploitation of all resources including cetaceans, for the benefit of man. For many centuries communication with animals was linked with witchcraft. In 1509, Cardinal Wolsey gave a banquet which included young dolphin on the menu. Dolphin meat was served in Parisian restaurants during the 'Age of Reason'.

The first account of a solitary dolphin episode I have been able to find, since the Classical period, took place in England in 1814. That year a four-metre adult male bottlenose, named Gabriel, made its home eight kilometres up the river Dart, at Stoke, in England. Gabriel was a favourite of the children, and adults, who watched his antics from the bank. He soon had a following of admirers from far and near. Then some enterprising showmen saw the chance to present a 'real live whale' in a sideshow at London's Haymarket. They netted Gabriel, laid him on a bed of straw in a large farm wagon and set off for London, exhibiting him in villages along the way. Before they had travelled many miles, the dolphin's liver and spleen were ruptured because of his unsupported body weight and he died in agony, expiring blood. The showmen eventually arrived in London with nothing but Gabriel's stuffed hide, which they exhibited anyway before finally selling the skeleton to a museum. Clearly the time was not ripe for interlock.

But by the end of the century, a transition was taking place. In New Zealand, for 24 years from 1888 until the early months of 1912, Pelorus Jack accompanied inter-island steamers along an eight-kilometre stretch of water between Pelorus Sound and French Pass. Its sex is uncertain but this was a beakless dolphin about four metres long, most likely a Risso's *Grampus griseus*. Besides bow-riding at speeds up to 15 knots, and rubbing its body against the ship's hull, there is not much to note in the behaviour of this dolphin except for the astonishing duration and regularity with which it met the steamers crossing Cook Strait between Wellington and Nelson, and its habit of joining and leaving them at a particular spot. There are some accounts of Pelorus Jack approaching open boats, diving to and fro beneath them and giving them a jolt, behaviour which parallels later Dints.

But the fame of this dolphin spread around the world, attracting thousands of tourists who would often make a double crossing just to see more of it.

By 1904 public concern for the safety of Pelorus Jack was sufficient to pressure the Government into legislating a special protective law, probably the first time in history a cetacean had received such recognition. No doubt, in some measure, this benign episode helped create positive attitudes towards dolphins in succeeding years.

Robert Stenuit presents the most unusual account of Sally Stone, a thirteen-year-old American girl who, on summer vacation in 1945, met a group of six dolphins in Long Island Sound, near New York. Sally liked to have herself towed behind a sailing boat. When the dolphins approached she showed no fear, even when they gently nudged her body. The next day she swam with them on either side, and one in front acting as guide. Day after day the games continued with jumps, circles and affectionate nuzzling. At the end of each session the dolphins would accompany the boat to port.

The following year Sally returned for her holidays and met the same six dolphins on the very first day. With embraces and caresses they resumed the gamesplay, towing Sally in circles as she held a dorsal fin.

In 1953, at a beach near Cape Town, South Africa, a pair of Indian Ocean bottlenose dolphins seemed to recapitulate aspects of the Hippo story. In March, two female dolphins began playing with bathers at Fish Hoek. According to C.K. Taylor,* he was able to make daily physical contact with both dolphins for many weeks. *Fish* and *Hoek* would tow people holding their dorsal fins and give rides astride their backs. The dolphins favoured a young girl, actively seeking her out from amongst the crowds of bathers. It was, however, impossible to approach the dolphins with any unusual objects intended for play such as rubber balls, pieces of rope or wood. They were quick to detect such objects and would move rapidly away.

Responses to Man of Captive & Free Ranging Cetaceans, Saayman G.S. and Taylor C.K.

75

Opo

Two years later, at Opononi, New Zealand, the best-documented Dint up to that time took place. A solitary bottlenose dolphin caused such a sensation that for 11 weeks she received regular newspaper coverage. To film Opo, three professional movie crews journeyed to New Zealand's remote north. Thousands of photos were taken and a great many were published. Especially impressive are several double-page glossy spreads in two competing weeklies. *Some 14,000 people were drawn to see this dolphin.*

As with many Dints, it is difficult to determine when people first perceived something unusual. During 1955, a solitary dolphin was interacting with boats on the sheltered estuarine waters of the Hokianga Harbour. Some reports suggest it may have begun the previous year when there had been a trio, including a smaller dolphin and a baby, and that following a shooting incident, the latter two disappeared.

A reliable account comes from a local Maori farmer, Piwai Toi, who tells in *Te Ao Hou* magazine how he first saw Opo:

'Although I had heard that there was a dolphin in the Hokianga Harbour I did not make her acquaintance until June of 1955. I was returning from Rangi Point school about 6.30pm and the sea was rather choppy. Suddenly there was a big splash and a boiling swirl. A large fish was streaking for my boat just under the surface. I really thought it was going to hit my boat, when about ten yards away, it dived and surfaced on the other side. It played round and round the boat. Such was the way I first met Opo. I was afraid she would be hit by my outboard, so I went inshore as close as I could. When I was in about four feet of water I looked back. She was about three feet out of the water, literally standing on her tail and looking at me from a distance of about 50 yards. She sank out of sight and that was the last I saw of her that afternoon.

'In August of the same year, two other chaps and I went to Rangi Point to gather pipi (shellfish). We had not gone far when we were joined by Opo. By this time, whenever we went out fishing, we were always on the lookout for her and rarely were we disappointed.

'Opo gave a charming display that day. She played round and round the boat and then swam just under the keel. When she did this you could feel the boat being lifted by the swell she made as she swam underneath it.

'One of the chaps sat right in the bow and kept putting his hand in the sea trying to touch Opo. At last he did. As far as I know he was the first person to touch Opo with his hand. While picking pipi three boats passed going to Opononi but she stayed just out from our boat cruising round. Then she followed us all the way to Opononi.'

Opo is the first Dint episode where it was possible to follow developments week by week in the daily press. In the absence of regular observations, newsclippings provide a useful record of the dolphin's gradual involvement with people. These can be compared with the story of Horace and others yet to come. I will present them here, in summary form, hoping they will convey incrementally some appreciation of the immense impact a solitary dolphin can have on human society; a response to that elusive quality I refer to as 'Ocean Mind'.

Dated 16 December 1955, the first news story is headlined: '**Opononi Has Dolphin Like Pelorus Jack** — A dolphin has been frequenting the Hokianga Harbour in the vicinity of Opononi for some months. First seen by locals after a heavy storm, it is assumed it took refuge in the harbour. Since then the dolphin has become a popular attraction in the area. It appears to make the Opononi wharf its headquarters and confines its activities to the waters between Hokianga Heads, about one mile south of Opononi, and a short distance north of the wharf. The dolphin escorts all small boats operating in the area, frolicking in and out of the water near them. When swimmers are in the water off Opononi beach the dolphin joins the party. Several people claim to have touched the "fish". Fears for the dolphin's safety are intense and the residents seek legal protection.'

On Friday 23 December, the first photos of Opo are published, showing her rubbing an oar, playing near a boat, cruising along the surface and waiting for the harbour launch. Now Opo is acting as a pilot between Opononi and Rawene, on the other shore. As soon as a launch moves she accompanies it. She is especially fond of rowing dinghies which she leaps alongside and passes back and forth beneath the keel, sometimes upside down. As she goes from side to side she gently bumps the oar with her dorsal fin, and will nuzzle an oar held just above the surface. Opo will go any distance up the harbour with a boat, but refuses to accompany even her favourite launch *Sonoma* across the bar at the harbour entrance. She meets outgoing boats on their return from the sea. When there are no dinghies or bathers to play with, Opo toys with a bottle in her mouth or balances driftwood on her beak. Some of her favourite locals occasionally touch her back. When displeased with people clutching roughly at her flippers she moves gently out of reach, smacking the water with her tail.

Perhaps Opo's most constant single playmate is 12-year-old Jill Baker, who enters the water five or six times a day to swim with her or to stand and talk to her. For Christmas, Jill receives a pair of bright blue flippers. As soon as she starts swimming up and down the beach Opo accompanies her.

By Wednesday, 28 December, the holiday crowds are the greatest Opononi has ever known. All campsites are occupied and icecream supplies exhausted. People are trying to touch Opo but she usually keeps just out of reach. Another of her favourites is Mrs Goodson, the local teacher, who swims with Opo each day. Now she is towed holding Opo's tail. Opo approaches a swimming spaniel dog and lifts it up several times. Strangely the dog is not upset.

Some New Year revellers on the harbour get a scare when a 'Loch Ness Monster' appears. Opo has seized a clump of floating kelp. With the stump upright in her mouth she sweeps along the surface, fingers of kelp streaming behind. She will also swim under a piece of wood drifting down with the tide, balancing it across her dorsal fin to show how fast and how far she can go before it slips off.

Early in the New Year, Opo is hit by the propeller of her favourite launch, *Sonoma*. While she receives two big scars, her friendliness is undaunted. Fears for her safety are regularly expressed in the press. An editorial on 4 January, calls her *'Opo'* for the first time, and urges special protection.

By 31 January, huge crowds are flocking to Opononi every weekend. Opo is performing from 8am until 7.30pm. When a boy seizes her dorsal fin and tries to lift her up, Opo reacts violently but then returns to play. Later in the day a boy has a tail-tow until dislodged. Children are playing with her in three rowboats. The dolphin rises beneath the keel and rocks the boat. As the children rub her belly with an oar the dolphin assumes a soporific, upright stance. She teases dogs by circling and then rising beneath them.

Fears for Opo's safety are growing. It is rumoured shots were fired at her some weeks back, and that she interferes with fishing nets. A vigilante committee is formed and notices posted: 'Please Don't Shoot our Gay Golphin'.

Piwai Toi describes the holiday crowds: 'One of my daughters who had to work at weekends as the proprietor of the tearooms was unable to cope with the crowds.

'By the time the Christmas holidays had begun Opo had really hit the headlines. The tearooms were doing a roaring trade and two other helpers had to be employed. We asked our daughter how many people were there each day and she said round about 1500 to 2000. We were a bit sceptical so we went to Opononi one Sunday afternoon, just out of idle curiosity.

'If I had not seen it for myself I would never have believed it. I have heard of traffic jams and crowded beaches but to see them at Opononi was a wonderful experience. This did not happen once or twice but every Saturday and Sunday. Cars, trucks, vans and motorbikes were seen parked on either side of the road for half a mile or more on each side of Opononi, with barely room to drive along the centre of the road. If a vehicle was held up or was to meet one coming the opposite way, a traffic jam was the immediate result. Traffic was so congested at times that officers had to be brought in to direct it. Two officers were on duty most Sundays and they did a very good job in untangling traffic. Two Sundays before Opo died, a special parking place was made available which was a real boon.

'With the record traffic on the roads I never heard of a single motoring accident in coming to or returning from Opononi. As for swimmers, there were easily over a hundred young and old in the water but there was not a single drowning fatality.

'With all these people coming during the weekends, Saturdays in particular, when up to 1500 people were jammed on the beach, there was no case of drunkenness, fights or arguments. Everybody was in the gayest of holiday moods.'

In the first week of February a British TV team came to film Opo's antics. She now allows some children to put their arms around her for short rides. Opo does not really like being manhandled and the gentlest of her friends receive the most favours.

The dolphin has initiated rides for Jill Baker, easing herself between Jill's widespread legs and carrying her along on her back for several yards. But Jill would never deliberately get on Opo's back.

The bottle game is evolving — Opo leaps out with a bottle in her jaws, flicks it in the air and then dives under to intercept it. She will not accept any bottle but invariably finds one on the bottom. One day this week she breaks her rule and accompanies *Sonoma* across the bar. Over the summer there have been two or three visitations by other dolphins to the harbour, on one occasion a school of about a dozen.

On Monday, 13 February, a news photo shows Opo upright, her side being massaged with a mop. At the weekend there are 1000 people at Opononi — 300 cars. On Sunday when a ball is tossed to her she ignores it, but on the Saturday when fewer people were present, Opo played with the ball for an hour, tossing it in the air or skidding it along the surface with her beak. When its owner recovered the ball, Opo dived for a bottle and continued the game.

On 15 February, a major photo spread appears in the national *Weekly News*. One picture is a classic: Mrs Goodson the teacher, waist-deep with Opo uplifted, cradled in her arms. Her face is a study in tenderness. The same day the rival *Auckland Star* presents eight photos headed 'Miracle Dolphin'. One shows a young man, Allan Wells, straddling Opo, ready to ride. In another, Ralph Stevenson, a boy, is riding on the dolphin's back in front of her dorsal fin.

On 20 February, the National Film Unit records Opo performing her full repertoire of tricks just four feet from the shore. Opo now plays with three balls alternately. When her friend Mr Williams takes two of them, Opo makes off with the third and Mr Williams pretends to chase her with the outboard. As well as dribbling a ball, Opo now punts it into the air with her beak, nudges it with her dorsal and then flicks it with her tail.

A pictorial spread on 22 February, in the national weekly, *Freelance*, shows Opo romping with a dog. Photographer Harold Paton, describes her antics: 'As soon as a dog gets in the water Opo goes like blazes, swamps the dog with a bow wave and then lies about as if laughing at the joke. She will swim slowly on her back, flippers out-stretched like arms with a beach ball balanced on her chin.'

Headlines on 23 February declare, 'Three Men Attacked Opo and Tried to Drag her up the Beach'. Fortunately one of the vigilantes intervened when the men had Opo in one foot of water. Just prior to the attack a woman had been seen fondling Opo (probably Mrs Goodson): 'Holding one hand on her back and the other under her belly, tickling with both, she would gently and slowly roll the dolphin over sideways two or three times, then let her rest a moment or two while she tickled her again before another series of rolls.'

Opo has a new ball game. She leaps on the ball, sinking it and then lets it shoot into the air. Despite public outcries, the Government can see no method of giving Opo legal protection.

News photos on Friday, 24 February show Opo pushing an inner tube. It was hoped she would leap through it but . . . The previous day she gave a ride to a four-year-old girl — the youngest yet.

The next day's paper shows Opo with a ball by her tail. The caption reads: 'Opo turns on her back, balances the ball on her nose and swims backwards. Opo also gets the ball under her chin, rolls it along her stomach until it reaches her tail. Then she dives, sending the ball flying out of the water.'

The reporter remarks on the special effect the dolphin has on the crowds: 'In some curious way Opo creates a bond between all watchers, young and old, Maori and Pakeha. Everybody feels better and more friendly for having seen her.'

Another writer describes the crowd response as 'mass pixilation'. People would plunge into the sea fully clad just to touch her satin-velvet skin.

By early March, public pressure for legal protection is strident. The Opo

Protection Committee seeks more authority. On Saturday, 3 March headlines announce: 'Opo Will Soon Have Full Protection of Law. On Thursday night (8 March) an Order in Council will be gazetted and will come into force at midnight. Opo is readily identifiable by scars caused when hit by a launch propeller.'

'Not So Playful' is the headline on 5 March. Over the weekend a uniformed member of the SPCA keeps an eye on Opo from an open boat as a crowd of 1000 watches. She frisks and frolics in two-feet shallows whenever the numerous outboards are silent, but some ugly tussles for possession take place, with four outboard boats vying to lure Opo away. Nevertheless from 9.30am to 4.30pm Opo loyally performs for the multitude and then escorts a fishing boat across to Rangi Point. Somehow she does not seem as full of beans as usual. At times oar wielders, instead of stroking her head, would jab the oar down hard.

A traffic officer controls the waterfront, directing taxis, trucks, cars and buses from all over Northland, and tourists from further afield. The worst stretches of the winding shingle road to Opononi have recently been widened.

Friday, 9 March, celebrated American travelogue-maker, James Fitzpatrick, arrives by flyingboat to film Opo. Despite four searches Opo cannot be found. A song about Opo has just been launched in Auckland by Crombie Murdoch and there is a new foxtrot dance — the 'Opo Roll'. Rumours are abroad that Opo may be dead even though the day before she was her usual playful self. At midnight last night the special legislation to protect Opo became law.

On 10 March 1956, the full front page of the news is devoted to Opo: *Opo, Opononi's Gay Dolphin is Dead.* 'Around midday she was found by an elderly Maori, Danny Boyce, while gathering mussels at low tide. Her injured body was wedged amongst the rocks at Kauere Point, five miles up harbour from Opononi.' A preliminary verdict that she was trapped by the falling tide is challenged later.

Over the ensuing days the press covers the funeral preparations. New Zealand is in mourning. Photos show Opo's shrouded form suspended from a tree with a loose group of sad people standing around. A pathetic figure on the beach is Jill Baker, Opo's special friend, who spends hours wandering silently along the sand where she played with Opo ever since Christmas. She was the dolphin's most devoted companion. The locals ascribe much of the dolphin's confidence in humans to Jill's gentleness.

Later accounts dispute the assertion that Opo was stranded. They ascribe her death to a gelignite explosion — possibly accidental. The human response to Opo's death is a milestone in the history of Dints. Possibly no animal has ever been accorded greater funeral honours.

In the afternoon of 12 March on the harbour shore, three men lower Opo into a nine-foot grave. An undertaker pronounces formal rites before a file of people representing various districts, Maori and Pakeha; each casts a ceremonial shovel of sand on the body. About 150 people watch with flowing tears as floral tributes from all over New Zealand are laid in the grave and telegrams from near and far are read aloud, including one from the Governor General and another from a prominent Maori tohunga (priest), requesting that Opo be buried alongside the Kupe memorial stone at Pakanae.

In subsequent years a fine statue of Opo with a boy on her back, the work of sculptor Russel Clark was erected on the Opononi waterfront. Many reminiscences were published, paying tribute to Opo, but the words of Piwai Toi are especially sensitive.

Opo Eulogy — 'She was really a children's playmate. Although she played with grown-ups she was really at her charming best with a crowd of children swimming and wading. I have seen her swimming amongst children almost begging to be petted. She had an uncanny knack of finding out those who were gentle among the young admirers, and of keeping away from the rougher elements. If they were all gentle then she would give her best. When playing with a rubber ball no one could help but be thrilled by her antics with it. She would push the ball along the water and then flip it in the air, catch it on her nose, then toss it in the air again. Or she would try and sink the ball by pressing it under her body or tail. She must have got it fairly deep at times as the ball bounced nearly four feet up in the air when it escaped from under her. Then she would toss it in the air and hit it with her tail. To watch her was one of the most fascinating sights imaginable.

'There was her game of playing with empty beer bottles. Toss her a beer bottle, empty or full, and she would treat it with disdain. She had to find her own bottle from the bottom of the sea. How she balanced it on her nose I cannot imagine but she would toss it quite a distance up in the air.

'I have tried to make her play with a glass ball, a float from a seine net, but she never seems to use it, which makes me think she could not see anything clear like glass. It had to be coloured before she would take notice. She even tried to toss a piece of brown paper which I threw overboard.

'Opo had a real weakness for the sound of an outboard motor. Many times I have gone to watch people playing on the beach with Opo. I would be about a quarter of a mile away with the motor idling when I would hear "oh" from her admirers on the beach. Next thing I would see Opo coming towards me. Many times I have rowed away from Opononi and started my motor, only to find Opo had left her admirers and was following my boat but I always returned her to the beach and then, by rowing a long way off before starting the motor, I could leave her to her friends.

'I have seen her following a boat which had its motor going full speed, yet she could overtake it without the least effort. Many times when she overtook a boat she would leap clean out of the water. She would hit the water and still, at top speed, keep swimming round and round the boat.

'One of the funniest sights I have seen was the crowd of amateur and press photographers trying to take a snap of Opo from a boat. Opo would surface on one side of the boat. By the time the cameras were focussed on the spot, Opo had dived out of sight. They would wait for her to appear in the same place. Instead she would appear in a totally different place. All hands would train their cameras on her again, but before a snap could be taken she would disappear. This would go on for a quarter of an hour or more. When Opo had tired of her teasing ways she would give a good pose for a perfect snap.

'People came to Opononi from many parts of the country, arriving in the morning and waiting for Opo to appear. She was nearly always handy, cruising around nearby. If an outboard boat was conveniently near, the owner was only too willing to go and get her. Once she heard the motor she followed just like a dog, playing and cruising round the boat. As soon as she arrived people swarmed to the wharf and the beach, taking snaps, marvelling, or just enjoying themselves watching her. In fact, I have felt sorry for her as she never seemed to have the time to feed during the day. If she had the urge to wander, an outboard had only to be started and she would return to her admirers again.

'Some people got so excited when they saw Opo that they went into the water fully clothed just to touch her. One chap was heard to say, "I didn't believe what I'd heard. Now I've seen Opo I still don't believe what I've seen!"

'Such was her popularity that I've seen the same people come weekend after weekend with their families to enjoy and to marvel.'

The Fish of Peace — To the Maori people the advent of Opo had a special significance, according to Piwai Toi and other elders. In 1950, on the point of a sharp bluff projecting into Hokianga Harbour, a three-metre obelisk of hard blue stone had been erected. There ensued a full scale re-enactment of the arrival of the great explorer, Kupe, at Hokianga one thousand years before. In AD 950, his canoe came ashore at Rangi Point, just inside North Head and then immediately crossed to the southern side, landing at Pakanae. Subsequently, Kupe is said to have circumnavigated New Zealand and returned to this same harbour before setting out on the long return voyage to Hawaiiki. The full name of the harbour is Hokianga-a-Kupe — 'Returning Place of Kupe'.

For the next five years, the Ngapuhi people laboured to inscribe the stone obelisk in memory of Kupe. Then on 18 March 1955, the biggest crowd ever seen in the district, more than one thousand Maori from throughout the country and a great many Europeans, assembled at Pakanae Marae. This was the unveiling ceremony for the memorial to Kupe, discoverer of New Zealand. Hohepa Heperi, a Maori elder, told Piwai Toi, 'Opo is the Fish of peace, a legacy from Kupe.'

Twelve months later, Toi notes, Opo was found dead on the rocks called 'Te Kauere-a-Kupe' and he concludes, 'These coincidences are certainly strange.'

Wallis

Across the Tasman Sea in another tidal estuary, a most controversial Dint ensued. Was this bottlenose Jo Jo, a male escapee from a dolphinarium to wander for two years along the New South Wales coast before approaching a human community? Or was it a pregnant female that came in from the sea and began accepting gifts of dead fish before giving birth? Whatever the truth, the story of Wallis is exceptional. The basic facts and photos cannot be denied.

One morning in November 1961, Lou Levy heard huffing sounds near his boatshed. Looking around he discovered a two-metre dolphin within the tottering wooden walls of the old Tuncurry baths. This was an abandoned swimming enclosure measuring 150 × 50 metres, extending from the estuary shore into 10 metres of water along its outer fence. With a rising tide swirling between the worm-eaten pilings, the newcomer was swimming up and down the length of the baths, possibly hunting for prey. Lou took some small fish from his boat and tossed one towards it. The dolphin seized the fish, turned it head-first in its mouth and swallowed. It then spun around and swam slowly up the baths, returning when it reached the end. As it approached, Lou threw another fish and this was repeated several times before Lou went off to work — to prepare his fleet of hire boats for the approaching holiday season.

At the end of the day, the dolphin was still circling within the baths, so Lou threw it more fish. To his surprise, the next morning it was still there. This time he let several fish drift down on the tide towards the dolphin and it ate them. Fishermen friends came to see the dolphin and gave him mullet and whiting. They watched as Lou splashed the fish on the surface before releasing them. The dolphin waited, head-on to the tide, for the food to come to it.

Each day it ventured a little closer until, on about the tenth day, it was accepting fish from Lou's fingers. Soon he was holding fish aloft and the dolphin was rising half its body length from the water to seize them.

Towards the end of December, holiday crowds began to arrive at Tuncurry. About 240 kilometres north of Sydney, this is a small resort town near the mouth of the Manning River, where the estuary expands into Wallis Lake. As more and more people heard about the visitor, Lou put up a sign advising that he would feed the dolphin, which he now called *Wally*, three times a day. Before long, the walkway around the enclosure collapsed under the pressure of the onlookers and Lou had to shift his routine to the beach side of the baths. This move delighted the children as the dolphin previously had been mostly in deeper water. Now it swam among them in the shallows and teased the local dogs by swimming in circles around them. Dozens of children came to be photographed by their parents as they fed the dolphin. With Lou, it was boisterous and playful, jumping up for fish, but with children it showed immense patience, coming in very slowly and gently, mouthing the offering. Within a week of his calling the dolphin *Wally*, it seemed to recognise the name and would approach and wait for a fish. He would call, 'Wally. Come on, Wally!' across the water to where it was rising for a breath. Immediately it would swim over, leaping several times, until it was alongside him in waist-deep water. Then it would lie against his leg, head-on to the tide. As he waded along the shore it accompanied him into the shallows where the children were awaiting. Lou would then hold a fish as high as he could reach and the dolphin launched itself from the water to take it. Lou would toss a fish far away and usually the dolphin would be under it as it fell, leaping to catch it. Sometimes it was Wallis who tossed a fish in the air repeatedly. Then it would turn, stand on its tail and swim backwards, to the applause of the onlookers. Lou noticed that the greater the applause, the longer it would maintain the act.

At times Lou tried to trick the dolphin by flinging his arm in a certain direction without releasing the fish. As Wallis swam off to get it, Lou would throw the fish in the other direction. Wallis seemed to sense this or could see the fish in the air, because it would change course and be under the fish as it fell.

In early January 1962, Lou began writing letters to the State authorities requesting permission to repair a portion of the baths enclosure so he could retain the dolphin permanently. The State premier replied, saying that the matter would be considered.

When the holiday season came to a close and the visitors dwindled to a mere few, Lou found himself spending more and more time alone with Wallis. They grew closer and he was able to stroke the dolphin. It would come alongside him while swimming so he could hold its dorsal fin and get a tow around the baths.

Eventually, the dolphin's fame led to a television presentation and all hell broke loose. At Tweed Heads, a bigger resort 480 kilometres to the north, Jack Evans, owner of the Coolangatta Aquarium, claimed he recognised the dolphin. It was a valuable male performer that had been swept from his dolphin pool during a cyclone two years earlier. From the ragged dorsal fin and aspects of behaviour he was sure it was his Jo Jo. Now a battle broke out in the media as the two coastal towns vied with fierce rivalry for possession. To thwart any abduction of 'their' dolphin, a Tuncurry vigilante group threatened to encircle Jack Evans' truck with 20 or 30 cars.

For Jack Evans, everything rested on the sex of the dolphin, which nobody was able to verify. Lou Levy's request for permission to confine the dolphin was firmly refused by the State authorities.

Then fate intervened. During an autumn storm one side of the enclosure collapsed and Wallis was discovered further up the lake. Despite Lou's enticements nothing would induce it to return to the damaged enclosure. For several weeks he visited it at various locations, further and further upstream from the lake, handing it fish and rubbing its beak. Then he lost contact.

On 29 April 1962, the *Sunday Herald* carried a headline: 'Dolphin Wally has a Baby'. Several people claimed to have seen Wallis out on the lake with a baby by her side. When fishermen called her by name she swam up to their boats, but that was all. Whether a male prison escapee or a pregnant female, this solitary dolphin had created quite a stir on the New South Wales coast and the hand-feeding situation was soon to have parallels on the other side of the continent, at Monkey Mia.

Nudgy

Half a world away on the Gulf Coast of northwest Florida, another resort community, Philip's Inlet, stands at the entrance to Powell Lake. Each winter this large, partially brackish body of water becomes landlocked. By springtime it is so swollen with rainwater that it blows its plug of sand, releasing a brown flood into the Gulf of Mexico. The area is popular with fishermen and several fishing camps, docks and other facilities support a small residential population.

After a hurricane in early September 1965, two bottlenose dolphins were found in the lake, possibly seeking refuge from the mountainous seas outside. A few days later the injured body of one was washed ashore.

The survivor, a near full-grown male, had to spend the winter in the lake but when it opened to the sea in spring, the dolphin remained, having developed companionship *with a dog*. At first this dog just stood on its owner's dock and barked its head off as the dolphin frolicked in front of it, but one day they began romping together in the shallows. Soon the local kids joined in, and over the summer friendship developed between the dolphin and residents of Philip's Inlet.

But not everybody was fond of the dolphin. Some of the fishermen resented the attention it gave to their boats when they were quietly angling in some favourite fishing nook. George Brown, one of the locals, had named the dolphin Nudgy after its habit of nudging boats with its beak until the outboard started. Then, with his beak close to the propeller, he would pursue the boat wherever it went. When it stopped, the dolphin tried to nudge it into action again. So many tourists complained about the dolphin's disruptive antics, bumping their boats and giving them 'shark' scares, that some of the fishing-camp owners decided this was bad for business. Many others argued that the dolphin was attracting people to the area. Controversy raged.

In spring 1966, some of those with fishing interests decided to get rid of the dolphin. When the lake opened to the sea they attempted to entice it through the entrance behind a fast boat. Repeated attempts with different boats failed. Nudgy would pursue them out to sea and then, no matter how hard they tried to elude him, the dolphin would follow them home again. Then somebody tried to spear Nudgy, leaving five prong marks on his smooth, grey flanks. Not even this aggression deterred the dolphin's friendliness. When an oceanarium heard of it and sent people to arrange capture, all the locals, even those opposed to Nudgy, joined forces against the captors. Now it was *their* dolphin and nobody was going to take it away.

One day Robert Burgess, a distinguished writer on marine life,[*] arrived to see Nudgy and photograph his antics. He saw children cavorting in the shallows with the dolphin. The instant an outboard started Nudgy streaked over and hovered behind the idling motor. When the engine moved into gear and the boat sped off, Nudgy seemed glued to the propeller with his beak. And when it stopped, Nudgy flipped the skeg up out of the water before

[*]*Secret Languages of the Sea* by Robert Burgess (Dodd, Mead & Co, 1982).

racing back to the kids. He turned on his back and encircled them, belly up. Then he righted himself and swam between a child's legs, surfacing in the midst of the group and letting the kids stroke and pat him for his fine display.

But conflict with the fishermen was not over. The next time Robert Burgess visited the area he got a shock. The fishing-camp owner whose dog and children first befriended Nudgy, had enclosed the dolphin in a wire pen so that he could not meddle with the fishermen. Each night he released Nudgy to feed out in the lake. Each morning, of his own accord, the dolphin returned to the pen. But the writer now found the dolphin subdued and wary, keeping his distance from visitors, avoiding the sides of the enclosure and squirting water at a kid who threw stones to get his attention.

Shortly afterwards Burgess received a note from his friend George Brown saying: 'We lost Nudgy last night.' He never asked what happened. He just hoped that one night when he went out to feed in the lake, Nudgy kept going out through the inlet to the open sea.

Nina

Around 1972, on the opposite side of the globe from Opononi, New Zealand, a Dint occurred which closely parallels the Opo episode (and Hippo) with intensive beach encounters, huge crowds and considerable social upheaval; but this time wetsuit-clad divers are an important element. In his book *Dolphins*, Jacques Cousteau presents the story of Nina.

La Corogna is a small seaport on the northwest coast of Spain. About 20 kilometres from there, at a spot called Lorbe Cove, a solitary female bottlenose began following fishing boats. The first person she approached closely was diver Luis Salleres, who worked the clam beds in the cove from his small boat. One day he found a dolphin watching him. The next day, when it ventured closer, Salleres tried to stroke it. To his surprise the dolphin accepted his contact with seeming pleasure. They became regular friends but the diver found that few people back in town would believe him.

Salleres invited Jose Vasquez, who was interested in marine biology and animal behaviour, out to meet the friendly dolphin. Vasquez watched as the diver descended. Soon a dolphin approached over the surface and dived, only to ascend shortly after with the diver at its side. Although a poor swimmer, Vasquez could not resist getting into the sea with the dolphin.

While Salleres was away getting his camera from the boat, Vasquez became paralysed in both legs with severe cramps. Terrified, he waved for help. 'At that instant, the dolphin, as though she understood what was happening, came very close and remained absolutely motionless in the water next to me so that I was able *to put my arms around her body.*'

With Nina supporting his weight, Vasquez had no fear and simply waited for Salleres to return from the boat.

One of Jacques Cousteau's diving team spent a week scuba-diving with Nina, documenting her behaviour on film. At each session she would usually appear on the surface within five minutes and spiral down the anchor line to meet Renoir on the bottom 14 metres below. She seemed to enjoy the line touching her body. When he saw her, Renoir would extend his hand and Nina immediately rubbed it with her genital area. The ensuing gamesplay would last 30-45 minutes and there was always a period when she would begin leaping in the air before returning to the diver for more frolics. Renoir felt that during this play the barrier between man and animal no longer existed and some strange understanding emerged. Other divers who met Nina were deeply affected in this way.

Cousteau describes the social upheaval that developed in La Corogna when Nina's fame spread from the locality and she became a Spanish national heroine. Following her appearance on television playing with the grandchildren of President Franco, tourists flocked to see her. She always maintained a base in the same area, about 300 metres out from the beach in Lorbe Cove. Showing no favouritism, she made a fuss of every boat that approached her. She mixed freely with all the bathers on the beach, allowing them to pet her, hold her tail and even ride on her back. As many as 2000 people thronged into the water to touch her.

Nina would never accept gifts of food. Each day, around noon, she would vanish for an hour, possibly to feed or rest. Once, when bottlenose dolphins came by, two entered the cove and spent 15 minutes with Nina. When they rejoined the school outside, Nina remained.

Each weekend the town was stricken with traffic jams. The price of land doubled and speculators moved in. Fishermen became tourist guides and prospered. The newspapers devoted full page spreads to Nina's exploits.

Concern for her safety grew and laws were passed for her protection. Outboards were forbidden in her vicinity. Fishing nets were banned from Lorbe Cove. Believing her tail to be vulnerable, noone was allowed to hold it.

In response to popular demand, newspapers launched a fund to build a 'Nina' monument on La Corogna jetty — a sculpture of a diver with his arms around a dolphin.

During the winter, some fishermen reported seeing Nina out near the clam beds. She seemed to be distressed. Five weeks later her body washed ashore along the coast. Human agency was suspected, but for five momentous months Nina had stood La Corogna on its head.

Charlie

The history of solitary dolphin encounters with humans takes a quantum leap once it begins to involve people who move freely beneath the surface.

In 1976, off the little fishing port of Eyemouth on the east coast of Scotland, scuba divers began meeting a large female bottlenose which they called Charlie, before its gender was known. Around 1960, this dolphin had frequented the coastline north and south of the Firth of Forth, near Edinburgh.

One day two Eyemouth diving club members, Archie Veitch and Ian Eaton, were 14 metres down, hunting for scrap metal on the wreck of the *President*. A huge, shark-like shape materialised beside them. They soon realised there was no danger and a cautious friendliness developed. Almost every time their club members visited the wreck, there was gamesplay with this dolphin. Each encounter revealed some new facet of her behaviour, according to a diver G.R. Mundey, who felt she was increasing her knowledge of human limitations underwater.* At the time he visited her with movie and still cameras, Charlie had established herself around a massive complex of rock slabs called Hurker's Rocks, which guard the entrance to Eyemouth harbour — a bottom of waving kelp forests, steep, current-washed rock faces adorned with soft corals called 'dead men's fingers', and a pair of great kelp-covered boilers, relics of a wreck.

When a massive grey shape appeared beside him, Mundey was surprised that he felt no fear. The gentleness of the creature was such that he had seldom felt more secure and at ease underwater, even though it was powerful, around three metres in length, and immensely broad with a heavily muscled tail. On the skin he noticed circular scars 'as if fingers had been pushed in to a depth of half an inch'. Such injuries are often observed on cetaceans and are usually the result of attacks by the small, deep-water 'cookie-cutter' shark which makes a practice of removing circular plugs of flesh and skin from its victims. These seldom lead to any loss of health and vigour.

Finding conditions near the wreck too murky for photography, the dive boat shifted to clearer water, out of the current. Charlie escorted them at high speed with upside-down swimming spurts which ended in a leap just ahead of the boat. When the divers descended and began making a variety of tooting sounds and 'Donald Duck' noises, Charlie went quite mad doing high speed, head-on charges straight for the divers' navels, only to swerve aside at the last moment.

Mundey believed the dolphin fed for only a few hours each day, around high water, when an especially strong current sweeps past Hurker's Rocks. This would happen at night too, since darkness would make little difference to a hunting dolphin. At such times, fish congregated in the sheltered lee of the rocks to avoid the tide race, and Charlie could be seen there breathing regularly and presumably feeding. On one occasion Mundey actually saw her snap up a couple of cod.

In the afternoons, when the Eyemouth fishing fleet of around 30 boats

*'Charlie Revisited' — G.R.Mundey, *Animals* 10, pp. 354-356.

returned to port, Charlie would escort each one into the shallows. But she would always abandon these vessels if the dive boat arrived and began dropping clumsy, slow-swimming playmates into the sea. To enhance their performance, the divers introduced an aquaplane to their games — a board on which a diver lies and can be towed quite fast behind a boat, regulating his depth with hand-controlled elevators. Charlie was delighted. She would station herself within a foot or two, rolling and turning alongside the towed diver for much longer than he could endure the effort of holding on, or the cold.

Mundey felt the dolphin had a clear grasp of the 'no win' rules of play. She obviously enjoyed being chased. Since it was not much fun streaking off on her own, she compensated by making slow and inefficient half-strokes, or moving her tail flukes in a series of tiny flicks so she was only fractionally faster than the diver. Then they looped, twisted and somersaulted a few feet above the kelp, his hand outstretched only inches from her flank, but never quite touching.

On another occasion she seemed to find new ways of slowing her speed by swimming head down with her flippers twisted to give maximum lift so that she had to force herself back — foremost through the water or swimming very slowly upside down. Mundey watched Andy, his acrobatic diving companion. Whatever this diver did, Charlie copied, even to the extent of lying flat on her back amongst the kelp, her great white belly uppermost, a rather undignified pose.

Mundey concludes that when a dolphinarium representative visited Eyemouth with a view to 'protecting' Charlie by catching her and adding her to his zoo, he got a cool reception from the people of Eyemouth. During the winter months Charlie disappeared.

Footnote: A recently discovered article tells of the initial period, to 1966, when Charlie spent 4-5 years around Elie, a hundred kilometres north of Eyemouth. Her closest friendship, with teenager Jane Cranston, involved swimming games and touch. Interaction with boats was so vigorous that she knocked the rudder off one and split the wooden hull of another with a tail thump. She leapt completely over boats, a canoe and a water ski tow-line, and she gave a sailing dinghy such a surging lift its speed increased markedly.

Donald

The next Dint to enter our history is quite epochal. The saga of Donald seems to unify many aspects of preceding episodes, while those that follow often recapitulate or underscore his exploits. This is the best documented and, until then, the longest term of human/solitary dolphin exchanges. None have been so peregrinatory, unfolding along such a long stretch of coastline.

In January 1972, biologists were relocating an experimental rig on the seabed near Port Erin Marine Laboratory, Isle of Man. Just before diving Mike Bates sighted a large bottlenose dolphin alongside their boat while he waited for a wire strop being lowered to lift the rig. Through the gloom, the dolphin appeared beside the wire. As the divers attached it, the dolphin stood on its head watching them, first with one eye, then the other. When Mike copied it, playfully pushing his mask within inches of its eye, the dolphin became excited, rushing around, charging in and veering off at the last moment. All the time they were shifting the rig to a new site, the dolphin kept them company and on successive days began meddling with their work, playfully. Donald's socialising had begun.

For the initial four years, Donald's movements and socialising were traced by Dr Horace Dobbs. Christine Lockyer, of the British Whale Research Unit, wrote an academic account up to December 1976, and Dr Nicholas Webb covered the final two years.

During this time the dolphin made a southward odyssey along some 480 kilometres of British coastline, from the Isle of Man to Wales, and then to Cornwall.

Along this course, he adopted a series of home territories for varying lengths of time, leaving in his wake reports of sociable behaviour. These viewed overall exhibit a remarkable development in variety and complexity.

The small harbours, boat havens and coves in which Donald stayed show similarities. Boating and fishing activities were intensive, with moored small craft, diving and swimming activities usually present. The bays were around ten metres or less deep, often with a rocky shore and sandy bottom. He had a fascination for wooden dinghies and mooring buoys, and would usually establish his 'home base' or territory around one or other, leaping over dinghies, lifting, tugging and circling buoys so that his whereabouts was soon known to the locals. If overcrowded in these 'special areas', he would become upset and apparently defensive.

Donald enjoyed yacht races, or contact with small craft and canoes, probably because of the opportunities they presented for interaction with people. He would push boats around, peer over gunwales at occupants, bite paddles, splash and capsize. The more excited people became, the greater his activity. He

Footnote: During an unexplained absence in 1975, when explosives were being used in Port St Mary, and before Donald arrived on the Welsh coast, I suspect he may have visited the east coast of Ireland where yachtsmen reported an episode with a friendly dolphin to whom they played classical music.

found many opportunities to interfere in the work of fishermen, meddling with their lobster pots and nets, towing their anchors, and preventing them from picking up mooring buoys.

Unlike Opo, he showed little interest, for the most part, in ball games, but he had that same devotion to children and delighted in teasing dogs. He also rescued a dachshund.

As long as people were gentle with him, he would be docile and compliant, allowing them to stroke his head, mouth and jaw while he remained motionless, occasionally shutting one eye as if in a stupor. He even let scientists measure his length accurately from beak to tail (11ft 10ins) but he did not like to have people impose their will on him in a pushy manner. Nor did he accept offers of dead fish. He clearly recognised certain individuals from voice or appearance and showed a marked preference for the company of sensitive, intuitive people. Donald often towed swimmers short distances while they held his dorsal fin. At times he became sexually aroused, engaging mooring ropes, buoys and small boats in masturbatory acts, as well as adult humans of both sexes — but never children. Towards some individuals, and certain dinghies, he displayed a possessiveness bordering on aggression but such episodes occurred at peak holiday season when Donald was crowded by more people and boat traffic than he normally experienced — situations which included extreme excitement.

As with Nudgy, outboard propellers held a fascination and he found he could make them cavitate by exhaling beneath the blades so that expanding bubbles altered the noise pitch — to the boatman's consternation. The rattle of anchor chains had equal fascination for Donald, and he showed interest in pulsed sounds and the whine of camera motors. He seemed to use his lower jaw as a sensor for exploring new objects and would gently mouth swimmers' legs and arms.

But the aspects of Donald's behaviour most relevant to our study are his attempts to communicate with the many people he met along the way. The rich vein of anecdotes he left in his wake show the resourcefulness with which this solitary dolphin used body language and the touch channels available to him, as well as the degree of empathy he displayed towards humans. Apart from a few weeks during his initial sojourn in the Isle of Man, when Donald was accompanied by a smaller bottlenose and avoided people and boats, for six years of his life this adult male dolphin was dependant on another species for the social interaction which is so important to his kind.

Horace Dobbs' books, *Follow a Wild Dolphin* and *Save the Dolphins*, serve to flesh out the excellent academic papers written by Lockyer and Webb.

On one occasion Dobbs lost his brand-new underwater camera. Searching in the murk he felt a playful nudge — Donald. Desperate to recover his precious possession and most reluctant to play, he spurned repeated nudges. Eventually the diver was induced to follow the dolphin until, standing on its head, flexing its great body like an arrow, it pointed at the lost camera.

When Dobbs tried to photograph Maura Mitchell, Donald's special Isle of Man friend, sitting on the seabed with the dolphin, Maura placed a rock in her lap for stability. Donald became so agitated at this he twice nudged the rock from her lap. When Maura swam around and sensitively demonstrated to the dolphin that she was in no danger of being trapped, Donald accepted the situation with tranquillity.

On another occasion, like Nina, Donald rescued a diver in distress, supporting him gently on the surface, helping to tow him to the boat, and remaining alongside until he had recovered. Yet, just prior to this, the dolphin had completely upset a diver-training session involving *simulated* rescue situations.

One day in Port St Mary, Isle of Man (October 1974), quite spontaneously Donald took Horace Dobbs' thirteen-year-old son, Ashley, for a ride around the harbour, depositing the delighted youngster in front of his father.

Dobbs' book gives detailed accounts of the subtle ways Donald would change a situation to gain his own ends.

While making the film *Ride a Wild Dolphin*, in Cornwall (June 1976), Dobbs was towed behind a boat on an aquaplane. Donald appeared as soon as the tow started: 'By this time,' writes Horace Dobbs, 'I was beginning to understand how Donald thought and I could read his signals. Thus when I went off at an amazing speed on the aquaplane I was not surprised to see Donald cruising easily alongside me. However, after a short time Donald made it clear to me that he wanted to play with the aquaplane. At first he gently butted me with his head, then he tried putting his head between my arms, forcing me off. I knew exactly what he wanted but decided that I would not concede to his wish. So Donald started to nip my elbow with his front teeth. Eventually he bit so hard I had to let go. Having got me off, he then tried to grab the aquaplane in his teeth, and get a tow himself. But the board was made of rigid, slippery plastic and he couldn't grip it. Even so he spent quite a while behind the boat trying to hang on to the aquaplane, while I climbed into another boat nursing my arm. When I rolled back the sleeve of my wetsuit I had a neat row of bleeding wounds inflicted by Donald's conical teeth.

'An interesting aspect of this behaviour is that both the dolphin and I were fully aware that he could have annihilated me in a trice by ramming me with his beak — in the same way he might have disposed of a shark. Yet he chose to gradually increase his pressure on me until I acceded to his wishes. This, I suggest, indicates a great deal of sensitivity on the part of the dolphin in his desire to communicate.'

In later months the towing game developed to the stage where Donald would tow another swimmer after the boat until level with the aquaplane, even while the boat followed a zigzag course (Webb, July 1977).

A second aspect of Donald's signal behaviour did not become apparent to Dobbs until he had seen their film several times: 'At the beginning of the film Donald is playing with Maura Mitchell. He then *hangs his tail over her head* and brings it down dramatically as the camera crew move in for a close-up. That was in Part I. I remember filming the beginning of Part II very clearly. We introduced Donald to a ball. At first he did not know what to do with it. Then he invented a game in which he flicked it high in the air and jumped over it. When I went to join in, he charged straight at me. For one terrifying moment I thought he was going to ram me. But he stopped short. I got the message clearly — he was playing with the ball and didn't want me to interfere. Although the camera did not film him charging towards me it did record his next gesture which was to *hang his tail over the ball* for some time. I now think that in doing so he was expressing to us that it was his ball and he didn't want us to interfere with his game. Likewise, when the camera crew had moved in to film Maura, he signified that he did not want them to interfere. When they subsequently went in closer, against his indicated wishes, he beat the inflatable with his tail and nearly tipped the cameraman and the director into the sea.'

Of the many anecdotes collected by Horace Dobbs, among my favourites is the story of Donald and the yacht race. On a summer evening at Dale Haven in Wales, a fleet of sailing dinghies was becalmed. A dolphin surfaced beside a boat and the skipper relieved his boredom by chatting to it. He was familiar with Donald's exploits and told him how pointless it was, a yacht race and no wind. The dolphin vanished. When his dinghy began to move, the yachtsman concentrated on his rig, intent on getting every tactical advantage from the least zephyr. Then he found he really had no control of the boat. It slowed and he heard the huff of the dolphin alongside. When he began to surge forward again, he realised the dolphin was assisting him. As a result, he won the race. A great debate developed at the yacht club as to whether the rules allowed for dolphin power. Matters worsened when it emerged that another boat had actually been pulled backwards. The argument is probably still raging in that Welsh club house.

Boat Towing — During the summer of 1977, while Donald was on the Cornish coast, he drew increasing attention to his presence by towing yachts. At this stage Dr Nicholas Webb took up the association with Donald. For three months that summer Webb followed the dolphin's travels around the coast of Cornwall, swimming with him and studying his behaviour.

The most intensive period of towing activity was a fortnight in August 1977, at the height of the holiday period, when Donald took up residence in Coverack Cove. A kilometre across, with a sandy bottom at about 14 metres deep and containing seven permanent moorings, the cove includes a small stone-walled harbour that dries at low tide. Over a nine-day period (August 4-12), *Donald towed every vessel* that anchored in the cove, including three schooners of five, six and seven tons. The largest schooner had a 35-pound anchor and chain, but Donald picked it up and towed the vessel 100 metres in a semi-circle.

A small cabin boat was towed 500 metres into the centre of the cove and eventually returned close to its original position. When three family sailing cruisers dropped anchor one evening, Donald towed all three. The first gave up and left the bay. The second, after being towed, tried to pick up a permanent mooring buoy but Donald repeatedly pushed the float just out of reach of the crew. (Such teasing was a common practice during his travels and a Falmouth skipper had been similarly frustrated for over an hour before giving up.)

The third of the sailing cruisers dropped anchor and the crew went ashore. Nearly two hours later, Donald picked up the anchor and towed the vessel 20 metres to a permanent mooring buoy. A diver, Colin Swaler, who was in the water, had been watching. When he grasped the anchor and chain to reposition the boat, a tug of war developed with Donald. Seeing the dolphin's jaws slide up the chain from the anchor, Colin dived down and wrapped the anchor around the permanent buoy rope as the dolphin watched. When Donald found he could no longer move the boat he swam to and fro between the diver and the tangled anchor, nodding his head at Colin. The yacht's skipper then returned and winched up the tangled mess while the dolphin pulled in resistance until it was clear of the water. He then swam at Colin, jaws open and gave him two sideswipes with his tail. Colin withdrew rapidly. This whole performance was confirmed by people on shore.

The next evening, a three-ton yawl dropped its 30-pound anchor and chain. Immediately, this was seized and the ensuing tow lasted two hours. Donald held the chain in his jaws but wrapped over his head to reduce the strain.

Photos show him actually exhaling through the chain links. The Dutch skipper was fascinated and did not interfere. He stood on deck taking photos. Donald was extremely excited throughout, several times leaping over two children swimming in the area and frightening them badly. Donald would not usually behave like this toward children. During the tow, three permanent mooring chains were entangled with the yacht's chain in a horrific snarl that took two days to fix. As the mess was winched up Donald watched, only inches from a concrete block as it hung at the surface.

The following day, a regatta took place in the cove but Donald did not tow *anything*. Three days later, during a storm, he left the area.

Following these observations, Webb received reports of other towing episodes that had occurred the previous month. Up at Helford, a ten-ton catamaran had been towed half a mile at speeds of up to two knots. To Webb the towing behaviour seemed to be a form of play demanding great effort and single-mindedness. Noticing how boatmen take possession of their boats, move them around and moor them, the dolphin, Webb felt, was capable of 'deferred imitation' or correct, first-time imitation of complex models. Such detailed, accurate, well-documented observations of dolphin behaviour as those published by Dr Nicholas Webb offer rare insights into their capacities.*

Dr Webb wrote another scientific paper about Donald's sexual behaviour toward woman, and his carrying of women and children which he terms 'possession'. Anybody studying this aspect would find parallels in the behaviour of the male *Tursiops*, Horace in New Zealand, and other solitary male dolphins in this story.

As far as is known, the first person Donald ever contacted physically was wetsuit clad, woman diver, Maura Mitchell, in the Isle of Man phase, 1972. Dobbs' book gives numerous instances of the lengthy and tender relationship between these two with croonings and cuddles at each meeting, while the dolphin shut its eyes.

When he met women subsequently, they were often bathers and not clad in neoprene. Women usually behave gently towards dolphins and are on a more receptive level than the diver-photographer who is eagerly posing for his friends. Just as Nina became sexually aroused when petted, so would Donald, and other solitary dolphins. Boisterous responses to unintentional wooing sometimes alarmed new acquaintances. With such people it seems, Donald did not repeat the behaviour.

Gentle mouthing of limbs was Donald's body language for establishing mutual trust (as with other Dints, both orca and dolphin). If the swimmer showed no fear, the dolphin usually stopped. But in cases where a person was not prepared for this, the dolphin continued the gesture. Resistance occasionally led to scratched legs.

Down in Cornwall in his sixth year of Dint, Donald gave unsolicited rides to people more and more frequently — men, women and children. Fear was sometimes aroused and such incidents could well be described as 'possession' by the dolphin. But on other occasions, as with Ashley Dobbs, such events were regarded with delight. In this respect, I see some ambiguity. Donald's public relations problem could be seen as a communication gap between himself and the people he met. He could *not* assume a pooled fund of experience, transmission and continuity in all those he approached. When people initiated play activities his responses overjoyed them, but when the initiator was Donald

*'Boat Towing by a Bottlenose Dolphin'. *Carnivore* Vol 1, Pt I 1978.

his intentions were sometimes regarded as ambivalent.

Thus, while in all his human encounters the dolphin was exceptionally benign, at times he really did give alarm. The apparent inconsistency might be resolved if each episode could be examined in the context of Donald's previous experiences, for herein lies the key to understanding human/dolphin relationships and the problems involved in meeting an alien mind.

Abduction and Dominance Behaviour — In some episodes of dolphin 'possession' there *is* sufficient material to provide a context in which the events took place, making them more understandable. In his fifth year of solo travel, Donald arrived in Coverack Cove in July 1977. For a month he had been developing a rapport with two women who took regular swimming exercise alone in the bay. During meetings with the dolphin, they would stroke and caress him. Then, in early August, a team of three male divers arrived in the cove with an inflatable boat, cameras and equipment — Drs Horace Dobbs and Nicholas Webb, and Ashley Dobbs — all eager to study the dolphin's responses to experiments they had devised. They had agreed that all their tests would be dependent on the dolphin's cooperation.

On the afternoon of 5 August, they engaged Donald in boisterous towing games using an aquaplane to initiate activity.* Within a few minutes of a tow, the dolphin would begin mouthing the diver's forearm with increasing pressure until, by the third or fourth squeeze, he had to release the board. The dolphin would then stay with the swimmer, giving him a dorsal-fin tow, or he would pursue the aquaplane. Sometimes the diver was towed alongside the aquaplane, even when the boat pursued a zigzag course. When the boat returned for the swimmer, Donald sometimes made abbreviated tail-swipe gestures towards it: 'Keep away.' Then he would seize the piece of chord attached to the aquaplane and drag on it.

The most dramatic episodes of possession took place in this context. Later that afternoon, Jennifer Nias (who had met and stroked Donald while in the area the previous day) was taking her usual swim 15 metres from the divers. During a dorsal-fin tow with Horace Dobbs, the dolphin broke off and headed towards Jennifer. In her own words:

'There was a snort beside me and the water swirled as I turned. Below me passed the familiar shape, silver belly turned towards me as it surged beneath me. He surfaced to the side of me and rolled and dipped more energetically than before, and this time with an erection. He disappeared and I swam towards the boat. Suddenly he appeared, swimming straight at me below the water, jaws agape; he took my leg and then my hand in his jaws. Even when I brought my leg up to the surface his jaw followed carefully, so as not to hurt, but sharp. Minutes afterwards, at intervals in the swirling boisterous play which followed, he made determined efforts to take my legs in his mouth, at one time having my thigh imprisoned between his teeth. At this point I was frightened to move too precipitously, in case I caused him to bite, or I injured myself.

'This was a time of twisting, surging physical contact. Continually he rubbed his belly and genitals against me, repeatedly moved, and I with him, on his back with his penis hooked around my arm. As we played, I became bolder, standing on his back and belly, pushing myself into the water, turning to anticipate his next assault.

*This coincides with the intensive boat-towing episodes.

Credit: Horace Dobbs

For seven years (1972-78) a male bottlenose called *Donald* played with people along the British coastline: Isle of Man, Wales, Cornwall. In a trust gesture *Donald* mouths the hand of Ashley Dobbs.

Credit: Horace Dobbs

One of the closest friendships was with a scuba diver, Maura Mitchell, who first met *Donald* in the Isle of Man, 1972.

PLATE 10A

Credit: Tricia Kirkm

In 1982 *Percy* established a similar pattern of interaction to *Donald*'s, centred around Godrevy lighthouse, Cornwall, whe he had his home base near a reef with food-rich tidal currents.

PLATE 10B

Credit: Mark Gutter

Ruth Wharram and Hanneke Boon in an intimate moment with *Percy*: at times gentle and tranquil, at others vigor and leaping.

Credit: Tricia Kirkman

cy meets with Horace Dobbs and the Wharram catamaran. Dobbs has made films and written books about encounters h six solitary dolphins.

PLATE 12

Credit: Rico Oldj

In 1984 *Simo* became established near a tiny port in Wales. He could be very possessive of female friends and sou ̣
to sequester them. With Carola Hepp, he pushes with his beak, belly up and penis erect.

'His play became rougher. He dipped and plunged closer to me, his powerful, sinuous body curving in the water like a taut bow. He surfaced so close to me now, and at such an acute angle, that I was often pushed sideways. Without warning he seemed to vanish and as I trod water, looking wonderingly round for traces of this dynamic beast, there was a shout from the boat. I turned in time to see the shape of his tail in a fan of spray. I felt the surge and rock of water, and the swirl as he passed beneath me. The next time I was facing him as he leapt from the water. The whole 3.6 metres of him magnified in my imagination to twice that size, crashed into the water beside me, missing me, so it felt, by inches. This was repeated again, and then he turned his attention to the boat, leaping out of the water just clear of the bows, dipping under it and stirring the water around it into a creamy foam. He seemed, at this point, to be using his tail to splash and display in a purposeful way that I had not seen before.

'By this time I was alarmed, not by any feeling that he would deliberately do me harm, but by the sheer power of that magnificent body and the apparent effortless energy with which he lifted himself from the water and turned within it. I was also certain that his attentions were focused on me again.

'Without warning he left the boat and appeared again beneath me. I prepared myself for more erotic rubbing or dangerous horseplay, and began to touch him with my feet, feeling for him with my hands. Instead I felt his head, or so I took it to be, underneath my stomach forcing me out of the water. Uneasily, though at the time I did not analyse this feeling, I slid away from him. Immediately he was there again, and this time, holding my body stretched transversely across his head, *began to propel me out to sea*. The speed must have been considerable, since I felt the shock of a bow wave against my body before he lifted me out of the water. The sensation was rather like riding a powerful horse, but much more intense, and more alarming. The sheer power behind that lifting push was awe-inspiring, and I was frightened by the irrational feeling that he was purposely carrying me off. At the same time, I still did not feel that he meant to do me any harm. There was a secret sense of pride in having been chosen by this dolphin for this dramatic ride. At the same time it was an action so clearly purposeful, so lacking in chance, that I panicked at the realisation that I was powerless, being thrust through the water on the head of an animal who made me feel clumsy and inept in his element. Also, he was heading out to sea, it was late and I was cold, and although the boat could have followed me, I was frightened, suddenly, of being taken away from my own kind, further into the sea.

'So I came off his nose. It was not easy, as he tried to keep me there, but eventually I tipped head first into the sea and swam towards the boat. I gladly went into it, then, and watched the dolphin leaping and diving for some minutes, perhaps, though with none of the deliberate force which had characterised his actions in the previous 15 minutes.'

Observing from the inflatable Nicholas Webb felt the dolphin had turned on their boat because it approached too close.* They saw Jennifer being carried well above water for a distance of about 15 metres, her body at right angles to the dolphin's path, her head, arms and legs thrown back.

Two days later Webb was swimming with the dolphin's other woman friend,

*'Women and Children Abducted by a Wild but Sociable Adult Male Bottlenose Dolphin.' *Carnivore*. Vol.1, Part II, 1978.

Marjorie King, trying with little success to interest him in an inflatable rubber ring. Donald began to mouth Marjorie's legs. When Webb approached, he received some abbreviated tail swipes, just firm enough to warn him off. The dolphin then returned to Marjorie carrying her off backwards while she choked in the bow wave. Protesting and distressed, she swam to a buoy and clung to it. Donald lay quietly beside her. Webb swam over and summoned the boat. As it drew near, the dolphin showed renewed interest in Marjorie's legs but was distracted by Webb. When she left the water, to Webb's surprise, the dolphin's mood changed completely. Donald followed him quietly and the diver had his close attention for the next half hour of experimentation.

In spite of these episodes, the women were not discouraged, and continued swimming in the cove. Their subsequent encounters were not violent.

For some time Nicholas Webb had been observing the dolphin's sexual behaviour, seeing it rub its genital areas against male and female swimmers, often with an erection. It would do likewise, even to ejaculation, with mooring ropes, buoys and the hulls of small dinghies, around which it often established a territory. Donald was once seen carrying a four-pound bass in his jaws, taking it to the bottom, releasing it, rubbing on it with an erection, taking the fish in his mouth again for a short distance and releasing it.

In his own early contacts with Donald, Nicholas Webb would remain passive as the dolphin rubbed its stomach and genital region against his wetsuit. Donald would lie upside down, quivering in the water, eyes partly or wholly closed. When the diver began to initiate by rubbing the dolphin's genital area, and dorsal and tail fluke, to his surprise, Donald's sexual behaviour terminated completely. Webb wondered whether he had unwittingly established a male dominance relationship with the dolphin, whereas with women the courtship pattern continued to develop, with teeth raking of limbs and uninvited rides.

Interestingly, the first person ever to *touch* Donald back in 1972 had unintentionally performed the classic dolphin foreplay gesture. Dobbs relates how two men and one woman snorkelled out to meet the dolphin. Donald swam slowly around each of the men, and then, around Maura Mitchell. When he stopped head-on to her, she reached out a gloved hand and stroked the dolphin gently under the jaw, talking to him as to a horse. The next time he swam past, Maura held out her leg *so the tip of her swim-fin stroked his abdomen*. For the next 30 minutes the dolphin stayed close to Maura and ignored the other two divers — not surprisingly, because such a gesture is typical of the initial courtship behaviour of dolphins.

Three days after Marjorie King's experience, two young children (a boy and a girl) joined Nicholas Webb in the water with Donald. When the dolphin mouthed the boy's leg, the children were frightened and clung to Nicholas as the dolphin milled around briefly, trying it seemed, to separate them. When Nicholas became upset enough to punch the dolphin, it became quiet, floating alongside the human group while Nicholas actually rested on its body. Comforted, the children recovered and began to stroke Donald who lay with one eye almost shut. As the children began to swim around freely again, Donald lifted the boy on his beak, much to the child's distress. A boat came to his aid while Donald sped around Nicholas, giving him a number of abbreviated tail swipes and almost knocking the camera from his hands. When Nicholas yelled angrily, Donald floated alongside him for a moment before speeding off across the bay with an inflatable toy.

A year earlier (1976) at St Ives, a 14-year-old boy had been sequestered for two hours by Donald, and was rescued by a lifeboat.

The previous month at Falmouth Harbour, Donald had been playing towing games with 16-year-old Robert Sutton. In one session the boy was given 15 dorsal-fin tows. It all began near some dinghy moorings where Robert, holding a rubber tyre, was being towed behind the family yacht. Donald began swimming to and fro beneath him, nudging his side, until the boy let go the tyre and held the dorsal fin. After each tow the dolphin would float quietly alongside the boy until its dorsal was grasped once more. The tow would start gently and then accelerate for 10-15 seconds until either the boy let go or the dolphin rolled over and descended. After half an hour, tired and cold, the boy began to swim back to their yacht hovering nearby. For a brief time Donald gently obstructed the boy by pushing him away from the yacht, before he let him leave the water.

Nicholas Webb had a similar experience of dominance as he tried to return to his inflatable after a lone encounter with Donald. Three times the dolphin floated passively across his path. When the diver reached the ladder, Donald made a brief open jaw gesture towards his arms and then left. In his accounts, Webb refers to the tail gesture as an 'abbreviated tail swipe' because he once experienced the full force of a tail strike which was so violent that it almost swamped his boat. Nothing remotely resembling this was ever directed towards swimmers.

Dobbs tells of a related episode with Cornish diver Bob Carswell back in 1976. Bob and his wife Hazel had a special friendship with Donald at that time. Typically, Donald liked to spend long periods with such friends, whereas with an unfamiliar group of divers, he might play for only a few minutes before leaving them.

One day Bob was 20 metres down, intent on gathering sea urchins as part of his living. Having been ignored for ten minutes, Donald swam over the diver and pinned him firmly to the bottom for a few seconds. Bob was not the least alarmed but felt it was the dolphin's response to his neglect.

Donald showed dominance and considerable hostility towards divers who carried knives or spearguns — on one occasion holding a diver's arm firmly in his jaws while he tried to remove the knife, but causing no harm.

Ken Dunstan was amongst the last divers ever to play with Donald. In February 1978, following a wreck dive with four men, and a period of gamesplay involving dorsal-fin tows, Donald tried to prevent Ken from returning to his boat. After an awesome show of dolphin power, Ken's seven-year-old daughter entered the water. Donald gently lifted his head out alongside her and let her stroke him.

Dominance behaviour towards children, such as described earlier, was most atypical. During Donald's coastal travels he allowed himself to be surrounded by beach crowds with kids patting and rubbing his stomach and tail while he floated passively on his back, flippers out of the water, tail flukes curled, besieged by tiny, screaming ecstatic humans. In such situations there were no gestures of possessiveness, tooth rakings or erections. At no time did Donald show sexual interest in children. It is significant perhaps that in many cases of possessiveness (also evident in similar situations with other solitary dolphins), there has been a group of people vying for his attention, as well as an individual with whom he has spent quiet interludes. It may be that abduction is a non-verbal statement about personal preferences in some situations, while in others it is an extension of towing games.

Nicholas Webb felt that the dolphin was clearly able to distinguish male and female humans, perhaps by sonar scrutiny of genitalia, and that he chose

99

to court females. From all available reports it seems this courtship was likely to be more assertive if human males were present. With solitary women, Donald appears to have been gentle. Perhaps some day, with more observation, we may develop a better understanding of interspecies etiquette.

During Donald's six-year coastal odyssey, he suffered a number of mishaps: a gunshot wound to the head, a gash from an outboard skeg, a stranding by the tide and a rescue with a mechanical digger, two days' entanglement with a buoy rope around his tail, and explosive charges detonated within his home range. Donald was last sighted at Falmouth, February 1978, just prior to the worst winter storm on record.

After reading everything on record about Donald, it occurs to me that much of his complex behaviour served to draw attention to his presence, and to elicit the companionship he craved. In this respect it often demonstrates surprising manipulative capacities, reason and memory. At times the dolphin may have been frustrated by the lack of creativity in those he met who were intent on observing or recording him on film. At other times people were quite possessive towards him, imposing their wills on the dolphin, which he seemed to resent.

An overall picture of his behaviour shows a development — from person to person and place to place. With new acquaintances his games were often brief. With familiar divers he played for long periods and resented being ignored.

Dictionaries define intelligence as 'the capacity to meet situations especially if new or unforeseen by a rapid and effective adjustment of behaviour' or 'alert, quick of mind, having intellect, endowed with the faculty of reason, communicative . . .'

Regardless of academic arguments, to people who meet them in the sea there can be no question as to the advanced capacities of dolphins. So many divers' anecdotes reflect this acceptance of their high degree of understanding.

Here is an illustration: Tom Treloar was diving on a reef in Prussia Cove at the southern tip of Cornwall. Visibility was poor and he got separated from his buddy, Bill Weddle. He decided to circle the reef to locate Bill. Suddenly, he felt *uneasy*. Looking about he saw a shape glide by on the edge of his vision — bluish grey and as big as a bus. He fought back panic and continued to search. Then he got the urge to look behind him. A metre away was Donald. Surprise and joy took the place of fear. Donald had a friendly mien and passed slowly and gently within feet of Tom, who followed, only to lose him. Donald returned, nodding his head like an excited dog. Tom stretched out his hand and touched the beak. He examined all the scars on his body — a propeller cut, a bullet hole . . .

They swam side by side, Donald eyeing him, as Tom thought, in a protective manner. Then suddenly, with a flick of his powerful tail, he sped out of sight and there in front of Tom was the anchor chain. Donald had brought him back to the boat. A trail of bubbles appeared, moving towards the anchor line. The dolphin had led Bill back, too.*

For many divers, encounters with Donald became the greatest day in their undersea lives. In May 1977, when Les Kodituwakku surfaced from a dive near Falmouth, he'd lost his scuba tank boot. Those on the dive boat were frolicking with a dolphin but his buddy, Malcolm Mister, offered to help search for the boot.

'Forget it,' Les shrugged. 'Let's play with the dolphin.'

As though he had heard the conversation, the dolphin disappeared below the surface and next moment reappeared with the lost tank boot around his beak. He practically dropped it in Malcolm's hands as if to say, 'Let's get on with the game.'*

*Both anecdotes from first person accounts published in *Diver* magazine.

Sandy

San Salvador Island is at the southeast end of the Bahamas chain, 620 kilometres east-southeast of Miami. Twenty kilometres long by eight wide, the island has one motel and a charterboat fleet. It is a first-class skindiving resort. In seven metres of water off Long Bay stands a monument to Columbus — historians believe the voyager first landed at this point.

During the later stages of Donald's travels in England, I heard of Sandy, another lone dolphin behaving similarly at San Salvador. For a time there were two Dints in progress on either side of the Atlantic and both dolphins withdrew within a few weeks of each other.

I drew up a special questionnaire and sent it to divers who had experience of Sandy. In response, divemaster Chris McLaughlin sent a superb set of transparencies and wrote this letter to me:

'I am always interested in information about human contact with dolphins in the wild, whether one or more dolphins are involved. I will try to describe Sandy and his involvement with divers on San Salvador Island. I first heard of big-game boats seeing a friendly dolphin playing in their bow wake in early 1976. The contacts became more frequent and on a few occasions he approached dive boats during that summer. Of course when we saw him, we stopped the boat and everyone jumped in the water. Sandy left immediately, but would sometimes come back and swim with the boat for a while.

'In October 1976, we anchored at Sandy Point dive site where he would hang out. He approached the divers underwater for about 20 minutes, but very much like a shark. He would stay ten or more metres away and retreat when divers followed. From October until December, Sandy remained in this general area and came closer and closer to divers underwater or snorkellers on the surface. It was not until February 1977 that I saw him allow anyone to actually touch him. Most of the time we went to Sandy Point he was there. He derived his name from this — "Sandy".

'No one knew if he was a male or a female at this stage, but by June 1977 it was quite obvious as he would get an erection and rub it against the diver's side and back. From June 1977 until 3 March 1978 when he disappeared, he was extremely friendly, playful, intelligent, warm and wonderful. We had an average of 50 people a week share in the experience of hugging and posing with this one lone dolphin. Almost everyone got to hold him at least once.

'I would estimate his age at four to nine years judging from photographs of other *Stenella plagiodon*. He was approximately two metres long and probably 70 to 100 kilograms. He was never taken out of the water and weighed nor was he measured with a tape. It's easy to judge his size from photographs with people. To my knowledge he never ate anything offered him and we tried everything.

'During his very friendly phase (ten months) he would usually single out one person to play with; that could be a diving guest or a divemaster depending on his mood. Until they got out of the water he would follow them continuously and rub and nudge them, looking for recognition.

'He would model for photographers and bite down on an "octopus" mouthpiece if it was offered. In other words, he was a "big ham". The strobe flash never seemed to bother him at all. He created such an atmosphere of excitement at the hotel amongst the guests, that no one cared about the diving on the "wall", which is excellent. Everyone wanted one thing — to see Sandy for the first time, or go back and see him again.

'From November 1977 until 1978 when he left, he began travelling up and down the lee side of the island and you could never be sure where he'd turn up.

'I made six night dives during which he showed up. Nothing is more exciting on a night dive than to be 25 metres deep concentrating on a photograph when a two-metre dolphin drops by and nudges you for attention. I had people who were so scared they climbed up the side of the boat instead of the ladder. Sandy had that streak of mischievousness so common in captive dolphins, not to mention in many people. If people got tired of petting him and swam off down the reef, he would open his mouth and hold them by the snorkel or mask until they renewed their attentions, which they always did.

'He has been photographed by most of the prominent still photographers in the United States and on 16mm and 35mm movie footage by Al Giddings, Smokey Roberts and Jack McKenney. The January 1978 edition of *Skindiver* featured him on the front cover, as did the April 1979 *National Geographic*. He was never tagged, blood sampled, or run through testing for hearing or sight. Everybody got to know him so personally that those ideas were never carried out.'

By the time Sandy had spent 18 months at the diving resort, his fame attracted divers from the United States by the plane load. The tropical deepwater location meant that this Dint was especially a diver/dolphin saga, with optimal conditions of clear, warm water and year-round good weather.

In all, some 2500 divers met Sandy. One underwater photo shows 11 photographers in the background each armed with sophisticated SLR camera and strobe. In such situations Sandy was hard put to find a playful contact.

Timothy O'Keefe, a visitor to the diving resort, wrote:

'On arrival at Sandy Point our skipper put the engines in neutral and revved them to make more noise. Suddenly a cigar-shaped shadow was off our bow. There was a spray of water followed by a quick roll of grey triangular dorsal fin. Sandy.

'Dave, already in full wetsuit, jumped into the water to greet his buddy. Of all the divemasters on San Salvador, Dave seems to have established a special relationship. Dave grabbed Sandy's dorsal and was towed around the boat. He let go, then grabbed Sandy's long thin beak and prised it apart to show us his teeth. Next Dave stuck his head through the jaws imitating a lion-tamer. Finally, he picked up Sandy's head and tossed him. Sandy responded by swimming around in front and slapping his tail centimetres from Dave's face, showering him. So Dave grabbed Sandy's tail only to find himself quickly dragged down and under. Dave spluttered back to the surface.

'When I'd finished my film Dave climbed back aboard and we both donned scuba. I jumped in the water, Dave handed me my camera and strobe, and down we floated. Sandy was busy harassing the other divers. He quickly learned our weak points. He tugged the mask of one diver, causing it to flood. Then he went over and pulled on the regulator hose of another. Neither action was done with any great force. He was using just enough strength to flood a mask, not to rip it off, as he easily could have done.

'Another of Sandy's favourite tricks was to swim away from a boat as though leaving. Concerned that he was tired of them, divers would quickly follow, swimming hard to keep up. After Sandy had lured them far enough to make it an uncomfortably long swim back, he returned to the boat to try and entice the next group away.

'It's almost as if Sandy's antics were part of a carefully rehearsed game. Yet he is the one who's devised all the tricks — without coaching from anyone . . .

'I'd been keeping my distance to film Dave and Sandy. Evidently Sandy decided I'd ignored him for too long. He came straight up to my camera and stopped just centimetres away, almost as if posing. I stayed motionless to film, not swimming. Sandy decided I was no fun. He reached down and gave my leg a solid thump with his nose. When I still refused to swim he gave me another. I reached out and stroked his side. He stayed there as I caressed him, running the flat of my hand along his firmly muscled, rubbery skin.

'Soon Sandy headed upward for air. I looked down at Dave, framed him in my camera and waited for Sandy to return. When he didn't, I lowered the camera, only to get a nudge in the shoulder. Sandy was back, playfully seeking more attention. It was then that I took hold of his dorsal and feeling like some god out of Greek mythology, went for my ride. It wasn't as fast as I'd visualised — we moved slowly through the water while I hung off to one side so my camera gear and belt buckles wouldn't scratch the dolphin. The ever present smile was only centimetres from my mask. I was entranced.'

In December 1977 at the end of Sandy's stay, Michael Voss was swimming up from 25 metres when he lost sight of the dolphin: 'Then I felt a knock on my right forearm as Sandy circled me and put his mouth around my arm. He was very gentle and we swam like this for several minutes while I occasionally stroked his belly with my left hand.'

Pat Selby, a woman diver, wrote to us sending pictures and impressions:

'Sandy was such a special experience in my life I cannot imagine anything to equal it. I first met Sandy in December 1976. He appeared at Sandy Point, leading the boat. When I went over the side with my camera, however, he disappeared. When the boat started up again he led it for a few miles and then withdrew.

'It was several months later that he approached snorkellers and they were able to touch him. Only certain people could touch him.

'In July 1977 I took Al Giddings down to meet Sandy. He has many excellent pictures and was enthralled with him. Sylvia Earle, Al and I went back in September 1977, and Sylvia's children came down also. Sandy loved children and would give them a ride with a hand resting gently on his dorsal fin.

'There are experiences too numerous to mention but I can point out certain characteristics. First of all, Sandy had a remarkable sense of humour. He was not averse to removing your face mask when you weren't looking. He loved to rub against fins and rubber wetsuits. Often he could be heard before being seen underwater. His staccato echo-locating clicks were clear, but a few people could not hear him. If I did not pay enough attention to him he did not hesitate to pull my hair or tap my head with his beak.

'He would rub his body against mine and then give me a ride while I held his dorsal. On one occasion he very cleverly drew me away for almost half a kilometre from the boat — a long swim back with no help from my friend. He would always tow me away from the boat — never towards it. But he seemed to sense each person's water ability and personality. He never played rough with those who could not handle it. He seemed to prefer warm, loving people. The feeling was always mutual. On one occasion he was trying to support a woman diver who belonged to a photographic class. She was in trouble on the surface and had to be towed in by Chris McLaughlin.

'Photographer Chris Adair would free dive to 35 metres, accompanied by Sandy. The dolphin loved this human who wasn't cheating with bubbles spouting from his mouth. Chris and Sandy had a very close relationship.

'One day in July 1977, Chris was free diving with Sandy and the dolphin took him deeper than usual. He pointed with his beak to the reef directly below. There was the cross Chris had been wearing around his neck. The chain had broken. We surmised Sandy had attempted to catch the cross as it fell because it was indented with several of his teeth marks.

'Sandy had several prominent scars, one on his dorsal fin and shortly before he disappeared, he was caught in the prop on the dive boat just above the tail fluke. He had healed well from that wound and I last saw him in January 1978. In March he disappeared while a research vessel was on the island. Whether he was taken by a collector, fell prey to a shark or just simply went off to find a mate, no one knows.'

Dobbie and Indah

By 1978, a basic pattern had emerged for solitary dolphin encounters with people. That year now appears as a pivotal point in the history of Dints. Our files show more widespread interaction during that 12 months than at any period before. Both the Donald and Sandy episodes terminated and Elsa and Horace commenced in New Zealand. Thirty kilometres away from Donald's last haunt in Cornwall, a major episode had begun near Cape Finisterre in Brittanny. Continuing as I write, the Jean-Louis Dint is the longest on record.

At this stage I find it difficult to maintain a chronological sequence. In some cases it is hard to determine when a solitary dolphin first began interacting with people. The most unusual and protracted of Dints, at Monkey Mia in West Australia, began with 'Old Charlie' some time around 1964, and like Jean-Louis, still continues.

From this point I will change course and present those Dints which are still continuing, at the end of this history.

Dobbie — If certain episodes of solitary dolphin interaction are relatively short-term they should not be underestimated. Proximity to humans is risky. With Dobbie in the Red Sea, interaction with humans exposed him to a trigger-happy population.

Around Eilat diving resort in 1979, a young male bottlenose began involving himself with undersea tourists. At Coral Island, near the wreck of a Greek freighter, three British divers were film-making when they noticed a lone dolphin watching their antics. One diver was struggling with an octopus that smothered his facemask during a modelling session. Then a dive boat full of tourists arrived. The dolphin made three spectacular leaps and the cameraman snatched a little footage.

When Horace Dobbs joined the team a week later, he was eager to get more dolphin film.* After a five-day search they met the dolphin in a channel between Coral Island and the mainland. During the ensuing film session, Horace tried to interest it in the same kinds of play objects and situations that had produced such dramatic responses from Donald: a rubber quoit attached to a towline, various water toys and underwater music. Nothing elicited the least interest. This dolphin was totally rapt in the antics of the divers themselves. Perhaps in their eagerness, the humans short-circuited stages of development, which with Donald occupied many months.

The Red Sea dolphin, later dubbed Dobbie, playfully bit at scuba exhaust bubbles and imitated the divers' movements, wagging a flipper in response to an elbow; poking his head out of the water like the diver; bobbing up and down with him vertically in an alternating see-saw pattern; swimming alongside him, both doing the dolphin-kick. But whenever the diver tried to touch him, the dolphin, typically, remained just beyond reach.

Interestingly, like the spinner dolphins I have seen in Hawaii and spotted

*Save the Dolphins, chap. 8.

dolphins in the Bahamas, Dobbie had a remora free-riding on his body. These sleek scavenger fish attach themselves for a free ride on a shark, manta ray or even, for short times, a diver. They cause no harm to the host but gain some advantages for themselves. In my experience dolphins do not seem to resent their presence, except for a vigorous headshake when one settles over an eye socket.

In all, Horace Dobbs and the crew were able to have a three-hour film session with Dobbie before he withdrew.

Some months later a biologist reported finding the body of a dolphin killed by rifle fire — the markings indicated it was Dobbie. In the interim he had given veteran divers at the resort the most remarkable experiences of their underwater careers.

Indah — The Australian continent and Tasmania are separated by Bass Strait. Midway across lies a cluster of islands called the Kent Group, a popular skindiving area with abundant marine life and clear water. In the summer of 1982, a solitary bottlenose dolphin started swimming with people around the various islands.

Having heard stories of these encounters, young Chris Hayward got up early on the morning of 7 March, to find a two-metre dolphin cruising close to the beach at East Cove, on Erith Island. With his sister and a male friend, Chris entered the water. All three wore snorkelling gear and Chris carried a Nikonos amphibious camera.

When they first got in, the dolphin swam around them slowly, holding its breath for short periods (Chris could hold his breath for longer), rolling on its back and swimming close beside them, but always just out of reach. Chris took 30 photos during their encounter, using available light and Ektachrome 400 film. Although they assumed the dolphin was a female, naming it *Indah*,★ the photos clearly show it was a young male.

Indah would roll so as to expose his underside to them and then swim along the sandy bottom five metres down, nuzzling a piece of drift kelp before dragging it along on his beak, fins or tail and then leaving it suspended in mid-water. A challenge.

'My sister then swam to take the kelp and instantly Indah followed, eyes fixed on the kelp. As soon as my sister released the kelp, Indah raced in, grabbed it and swam off to leave it in mid-water again. I moved in, took the kelp and swam off. Indah cruised alongside watching my every move. Then as soon as I released the kelp, he shot up, grabbed it and was off with three snorkellers in hot pursuit.

'We played this game for over half an hour with Indah taking the kelp further away each time and us finding it harder to keep up. Indah finally swam completely out of sight with the kelp and the others left the water. I swam on to find him but as soon as I got close he grabbed the kelp and swam out of sight again.

'I followed and eventually caught up, only to see him seize the kelp and swim away. I could no longer follow. I had already swum the complete length of the beach. The dolphin had beaten me. He had the kelp and had won the game so I got out of the water. Indah had certain scratches on his right side that could be used to recognise him again. The last reported sighting was March 1983.'

★Indonesian: beautiful.

Percy

'Hullo — Donald's back!'

At Portreath, north Cornwall, in June 1982, professional diver Keith Pope found his work being observed by a lone dolphin. For a moment he thought it was Donald, who had often watched his activities in the Falmouth area during 1977. The dolphin came closer and Keith decided it was not his old friend because he did not have Donald's scars. So began another lone dolphin saga.

The story of Percy extends over three summers and shows a gradual development of intimacy which peaked with quite extraordinary behaviours, just prior to his withdrawal. Percy showed so many of the traits which Donald and other solitary male dolphins have manifested, that we are able to gain some idea of what may be general dolphin responses to humans, and what may be idiosyncrasies.

With Percy, things started slowly. For the first year people of Portreath gradually became aware that a lone dolphin was frequenting the area around Godrevy lighthouse. Between this island and the coast, near a rock where the current was the strongest, Percy fed most often.

The following summer, local guesthouse proprietors Bob and July Holborn, began meeting Percy on their inflatable-boat excursions. Homeward bound after their first outing in early spring (1983), they were startled when right beside the boat the dolphin made a great curving leap and took up position on their bow, thumping the hull with powerful upstrokes of his tail. When they hove to and drifted for half an hour the dolphin came close, on his side, looking up at them. Percy then accompanied them all the way home, peeling off when a hundred metres from shore.

From that time, Bob made regular excursions, taking friends and children out all summer to meet the friendly dolphin.

Eventually Percy would allow Bob to touch him, but only from the boat. Bob began towing a car inner tube. If he stopped the outboard, Percy would nudge the tube, toss it in the air or flip it vertically on the water. Whenever Bob dived with him, Percy would watch his every movement, especially his flippers, but he would *not* permit Bob to touch him when he was out of the boat.

During the 1983 summer, more and more people heard of the dolphin, made special efforts to meet him and sent us accounts of their experiences, including London students Andrew Crofts and Caroline Tombs. They had already met Jean-Louis, the female bottlenose that had been interacting with people in Brittany since 1978. (This Dint still continues, so it is described at the end of Part Two.)

In July 1983 Caroline and Andrew wrote:

'We have just returned from a weekend trip to Cornwall, where we saw Percy. We paddled our surf skis out to the island where he hangs out, near one of Donald's old playgrounds. A thick mist surrounded us, reducing visibility to ten metres, and a tidal race was running in the gap between the island

108

and the headland. Worried about our precarious position, I was overjoyed when Percy appeared beneath me. After swimming among us for a while he treated us to an amazing display of acrobatics, leaping high, flipping and twisting simultaneously, twice in a row.

'In our brief encounter we saw a dolphin very different in character to Jean-Louis — much more boisterous, and immediately familiar with us. As we paddled along, he swam with his dorsal touching the bow whereas Jean-Louis would have been content to swim alongside.'

In late August 1983, Bob Holborn noticed Percy was not acting normally. He was less playful and rather aloof. Then Bob saw him doing a series of backward flips into a strong current, cracking his head on the water.

Horace Dobbs arrived and discovered the problem — Percy had a large fish hook, with considerable line attached, embedded near his left eye. They could do nothing to help him and it was some time before it disappeared.

Ruth and James Wharram, who designed the catamaran Jan and I use for dolphin research, have been equally addicted to dolphins ever since they met Donald. At their boat-building base at Truro in South Cornwall, they had set up a research team called the *Dolphin Link* Project.

Ruth wrote to us:

'A couple of days ago fishermen anchored in the bay to pick up some lobster pots. Before hauling the last one, they decided to have lunch. They got out their sandwiches and began to eat. Percy apparently watched this with growing impatience. He went down, picked up the pot, knocked on the boat hull and presented the pot to the fishermen.'

Over the summer of 1984, Percy lost all fear of humans. Perhaps this began with the trusting relationship he had developed with Bob Holborn. Bob started trailing a foot in the water from his inflatable. In due course he jumped from his boat and lay floating on his back for a considerable time until Percy approached. 'He came right up to me, opened his beak and mouthed me from the tips of my flippers to the top of my head. When he seemed satisfied he lay his head across my chest and I stroked him.'

From that time a close bond ensued. But Percy began treating others with the same familiarity, chasing windsurfers and tossing them off their boards; teasing waterskiers and buffeting small boats. Bob began to worry that this boisterous behaviour might result in somebody being hurt and then retaliating.

As well as these activities, Percy began to show possessiveness, and sexual arousal similar to that of Donald. Val Owen went out to meet Percy in Bob's inflatable. She got into very cold water without the insulation of a wetsuit. Within seconds she was met by the dolphin:

'Wow, was he *big*! At first I was frightened but once I realised how gentle he was, I recovered and began to talk to him and stroke him. He was very curious about me, looking me up and down and jostling me about in the water. Then he started swimming under and alongside me. He made such an upheaval it was difficult to stay still. I swam away from the boat and stroked him each time he passed by. After ten minutes I decided to swim back to the boat but Percy was enjoying his game of bobbing me around in the water. Each time I started towards the boat he swam between me and it, and his turbulence pushed me back to where I started. It took several attempts before I got to the side.'

In the high summer of 1984, the Wharram *Dolphin Link* team, initially guided by Bob Holborn, began a series of visits to Percy with the purpose of gaining an understanding of his capacity for relationships with humans. Regular diaries were kept by all who participated. There is space here to present two such episodes.

On 1 July, sea conditions were slight with a gentle northeast wind and good visibility as the Wharrams' seven-metre Polynesian catamaran *Tiki* sailed out from Portreath, to be joined enroute by Percy. At midday she anchored and James Wharram entered the water.* He saw Percy standing on his head by the anchor, ten metres below. The dolphin swam around him just beyond reach in the same friendly fashion as the previous year. Hanneke joined James and they showed Percy bodily contact between themselves, demonstrating their relationship. Andre entered the water naked. Percy closely inspected Andre and at one point grazed against his legs. Alistair entered next. He dived deep and the dolphin copied him. Percy played generally among the swimmers. Then Alistair remained alone and Percy responded to his exceptional diving abilities. While Percy was looking at the anchor Alistair swam down to meet him. Halfway down, Alistair did a backward roll and Percy joined in the manoeuvre so both were rolling in unison. Percy started nudging Alistair's fins when they were still; his beak remained shut. Alistair noted a fungal infection on Percy's head and a sore on his lower jaw.

Hanneke re-entered the water, less boisterous now but diving down several times. With every descent Percy accompanied her. She would extend her arms to him. He would spin around, facing her closely and emit sharp click trains towards her. As Hanneke held the anchor rope, her head above water, Percy would nudge her ankles, then her knees and once he put his beak between her thighs.

James joined Hanneke; holding hands and rubbing bodies they revolved several times. Percy led them in circles. Then he slid his beak up the outside of James' leg and nudged him firmly in the chest. James thought this was a signal to leave the water as he was tired and breathing heavily. From what ensued he later reconsidered this.

Now in a wetsuit, Andre entered and James left. Andre remained on the surface with Percy directly in front of him for most of the time, staring closely at him, his head often breaking the surface. Then Percy began nudging Andre, starting at his feet and gradually working up the body. He even propelled Andre a distance of three metres. The nudges were firm. Percy put his beak behind Andre's knees in an attempt to spin him around. Then the dolphin spiralled down the anchor rope and began romping about as he ascended. As Andre held the rope Percy rested his beak on his arm. It was noticed that Percy continued nudging Andre's legs even while they were moving. Most of Andre's contact with the dolphin was this nudging.

Alistair was next in the water and met with spectacular leaps and turns. Obviously Percy responded quite differently to Alistair's enthusiastic diving and swimming, being far more playful with him than with anyone else. At one point, Alistair stroked Percy all along his body but he had the impression that the dolphin did not particularly like it. His comment afterwards: 'I am totally absorbed by Percy; completely oblivious of anything else while swimming with him. He seems to elicit vitality from me, urging me to do more than I thought myself capable.'

*All divers wearing wetsuits, unless stated otherwise.

For nearly four hours that afternoon, Percy showed his interest in every person who entered the water, swimming up to each one and regarding them closely. He did not, however, go over to the boys diving off the rocks near the beach. When his favourite fishing boat came by, Percy swam over, made a single bow leap and returned at once. At one stage he rested his beak on the deck of *Tiki*. When the catamaran set sail for Portreath, Percy accompanied her for 20 minutes.

On 18 July, there was scarcely any wind at all. As *Tiki* motored out from Portreath a slight swell creased the ocean, traces of fog — a perfect day. Around midday, with Percy already in attendance, *Tiki* dropped anchor. First into the water were Ruth and James. Percy showed immediate interest in Ruth's genitals. Although not encouraged, he was just as boisterous as the previous trip, bumping people forcefully with his beak; affectionately resting his beak on their shoulders as they stroked him.

When James and Hanneke showed mutual affection, strokings and holding each other, Percy tried to join in, pushing between them.

Later, when Hanneke was in the water naked, Percy showed no arousal as she stroked his stomach. However, when Ruth went in, wearing her full wetsuit, Percy showed a clear sexual display towards her, rubbing his stomach along her body and touching her with his penis. Ruth did not encourage him.

Seeing this paradoxical behaviour, Hanneke re-entered the water wearing only her wetsuit top. Now Percy was highly aroused and simulated the mating posture with her as she floated face down. He would slide up from behind her, belly up, until his flippers were level with her arms. This is the only position in which he would ever allow contact with his flippers. His penis would be between her legs at knee level. He would then turn sideways and hook his penis behind her knee. As soon as there was any firm contact with Hanneke's body he would start thrusting. All the time Hanneke was in the water, he persisted with this behaviour. When she was vertical he would approach likewise and rest his beak on her shoulder as she embraced him.

At 4.30pm *Tiki* left the bay, Percy following for ten minutes with the usual display of farewell leaps.

On six subsequent trips, the *Dolphin Link* team experienced several new behaviours. On 19 July, Percy made a spectacular series of leaps as the *Tiki* departed. A week later he accompanied the catamaran all the way back to Portreath Harbour where it seems he tried to induce the crew into the water by splashing them. Earlier that day, when two divers had joined him, Percy approached one, put his beak beneath his arm and pushed him to the surface. Later, on 13 August, he actually restrained a very confident snorkeller on his return to the surface. On this final trip Percy appeared to have undergone a mood change and was emotionally withdrawn and confused. His eyes were usually half closed but this time they were wide open and he wasn't as relaxed as usual. There were, however, frequent episodes of belly rubbing on swimmers and boats. He allowed his dorsal to be held and placed his beak against masks. He showed delight in the novelty of an outrigger canoe; its occupant was aided in climbing back aboard by being lifted up, standing on the dolphin's back.

Approaching a plastic surfboard, Percy rubbed his belly on the underside, rested his beak on the rider's legs and subsequently knocked the rider off, after staring at him with jaws agape.

From Gothenburg, Sweden, we received an account from wildlife writer

Lars Lofgren (13 August 1984):

'We went out with Bob Holborn in his inflatable. After half an hour speeding along the coast, wondering if we would meet Percy, suddenly he appeared beside us. For the rest of the trip to a sheltered bay, Percy kept bumping forcefully at the sides of the boat and the bow, as if trying to alter our course. As soon as we had anchored, people leapt in and Percy showed sexual arousal, hooking his penis on to their legs and sometimes thrusting. A girl said that if she had not been wearing a bathing costume he would probably have entered her. He hooked his penis behind my knee and began thrusting. I never saw any ejaculation. He persisted like this all day. It eventually became clear to me that he behaved like this when you remained immobile, floating on the surface. When you swam vigorously or when you dived he did not respond that way. Twice he towed me through the water as I held his flippers, belly to belly beneath him. When we were about to leave Percy dived down, took our anchor in his beak, lifted it to one side of the boat and began hauling in the anchor rope.

'With regard to his sexual behaviour it must be remembered it was his third summer (1984) of interaction with people before this began. Prior to this he would not permit people to touch him. Just a month before I arrived he began to show sexual responses. He also stroked his penis on boats. To me it seemed like sensuous play. Dolphins are well known as being erotic, and all lone dolphins who seek human companionship have eventually shown sexual responses towards people — even males.

'Percy is constantly inventing new games and clearly he grows tired of them after a while. Once I had the impression he was testing my swimming ability. He swam beside me, slowly increasing his speed as I tried to match him, whereupon he very slowly increased speed and so on, until I could swim no faster.

'Bob Holborn claims that Percy and he have developed a gestural language with up to 30 signs, but he says experiments and more intimate interactions are more likely when few people are present.

'But this does not mean close interactions were impossible with many people in the water. One day I saw Percy playing with swimmers for hours and hours out from Portreath Harbour. It was a perfect day with lots of people bathing. It seems Percy was afraid to enter shallow water so people had to swim out to him. (Perhaps Percy realised he had the upperhand so long as people were swimming — author.) Percy swam from person to person, playing continually. The water was cold and nobody could stay with him for long. He never seemed to tire and was sexually active the whole day. He was inventing new games all the time, such as leaping ahead of a canoe, and shoving it off course.'

Over the summer people's reactions to Percy's sexual behaviour varied enormously. Offended by his erection, one man tried to make him behave, with a hard push and words of reprimand. He received a firm bite from Percy. Conflict reached a peak as more and more people argued about his sexual responses. A girl who had swum nude with him quite candidly said that, while she had not encouraged intercourse, she would be open if the dolphin initiated sexual foreplay. Some local people were quite upset and legal action was threatened if such conduct were to arise. Tempers flared and voices were raised.

'Would you let somebody sexually abuse the dolphin?'

Credit: Horace Dobbs

obbie, the solitary Red Sea bottlenose, who met Horace Dobbs during a filming expedition, and was shot shortly afterwards.

Credit: Horace Dobbs

PLATE 14

Credit: Chris McLaug...

Sandy, a juvenile spotted dolphin, *Stenella plagiodon*, interacted with divers at San Salvador Island, Bahamas, 1976-78.

olling upside down in a tranquil exchange, or accompanying his favourite diver, Chris Adair, on a mission (below), *ndy* craved human companionship.

PLATE 16A   Credit: Chris Heyward   PLATE 16B   Credit: Chris Heywa

Indah, the solitary Bass Strait dolphin, extends a challenging version of the seaweed game to humans, 1982.

PLATE 16C   Credit: Chris Hey

'That's buggery.'

'Was this an aberrant dolphin or was such sexual behaviour normal?'

A conciliator said, 'If we were in Polynesia we could not do anything that offended the culture of those people. So in Cornwall, maybe we should wear costumes and not offend the locals either . . .'

During 1984, cetacean scientist Christina Lockyer made a study of Percy — his home range, swimming patterns and interactions with people. Her paper* notes several traits he shared with Donald and other solitary dolphins: towing boats, nuzzling propellers and involving himself with fishermen's activities. On one occasion he deliberately entangled the buoy lines of Bob Holborn's fish pots. When Bob dived to sort them out, Percy indicated the reverse sequence with his jaws so they could be disentangled without being cut. The scientist remarks on the problem-solving capacity and memory this would involve.

She offers instances of Percy selectively pushing people from his territory — head butting, hard enough to cause bruising; abduction; biting, to draw blood; and other forceful aggression. In some cases his behaviour was not understood because, for the most part, he was under considerable social pressure. One surprising episode involved Percy swimming belly up to urinate over the stern of a boat occupied by startled occupants.

She ascribes his altered mood and sexual activity to spring and autumn peaks, behavioural patterns noted by other scientists. Her study of tooth-raking marks on his skin indicated that Percy occasionally met up with other dolphins and could not be regarded as truly solitary.

The final words on Percy go to Horace Dobbs, who was one of the first to make contact with him underwater:

'I watched his association with humans progress from swift passes at the limit of visibility to a stage where he was aggressively masculine and assertive. This final phase started when we both discovered that he could push me almost out of the water if I put the ball of my foot on his beak when he came up beneath me. This eventually developed into extremely rumbustuous play during which he would jump over me and tow me through the water at speed.

'However, just before *his final appearance at the end of 1984*, he showed that he could be exquisitely sensitive. It happened shortly after I had had a particularly vigorous game with him in which there was a genuine inter-species rough and tumble. I persuaded Tricia Kirkman, a woman who could not swim a stroke, to get into deep water. She wore a wetsuit which I assured her would keep her afloat, no matter what happened. She was terrified but had an overwhelming desire to get in with Percy.

'When she did, he calmed down immediately. She put her hands down and he allowed her to rest them on his head. Then, very very gently, he moved in a circle around our inflatable, remaining at just the right depth to tow his passenger smoothly on a ride which calmed her trembling body.'**

*'The history and behaviour of a wild, sociable bottlenose dolphin (*Tursiops truncatus*) off the north coast of Cornwall', *Aquatic Mammals* 1986, 12.1,3-16.

**The Tale of Two Dolphins* (Jonathan Cape, 1986).

The Costa Rican

News does not travel rapidly between Central America and New Zealand, but one day I received a newsclip from *La Nacion* (21 January 1983), with an extraordinary photo of a teenage boy standing in a boat hoisting the tail of a large bottlenose dolphin to waist height. The dolphin's body is curved through 90 degrees so that its submerged head is alongside the hull, its dorsal level with the gunwale. For a dolphin to voluntarily place itself in such a position, its capacity to breathe entrusted to a young human male, fairly took my breath away.

Gradually, I managed to glean further details. The people of Chira Island off the Pacific coast of Costa Rica, say that the dolphin came to their village after its companion had been shot by fishermen near another village. The dolphin is said to have followed a fishing boat back to their village. Friendship developed with young and old alike, but initial interaction was with a playful dog.

When news of the dolphin reached staff of the World Society for the Protection of Animals, Costa Rica based field-officer Gerardo Huertas was sent to check it out. Huertas found that the three-metre male bottlenose was making daily visits to the village and remaining in the vicinity throughout most of the daylight hours. It could be called close to shore by thumping on the side of a wooden dugout canoe with the paddle. As soon as people began swimming out from the beach the dolphin usually appeared. It would play with a variety of objects and often pushed around tiny dugout canoes with very small children in them. As a result, its beak was constantly chafed and raw at the tip. Huertas visited the dolphin several times and spent hours swimming with it.

When his reports reached WSPA regional office, Boston, director John Walsh decided to see the dolphin for himself. In response to my inquiries (I sent him a standardised questionnaire form) John wrote the following account:

'During the time I spent with the dolphin, it allowed me to get on its back, holding its dorsal fin, and would gently swim close to the surface of the water, not at any time trying to take me under. It enjoyed playing with children, often following them right up onto the shore, requiring help by pushing to get back into the water. It would allow a group of people to pick it up and hold it out of the water for short periods of time, and when it wished to return, would simply roll out of the arms that were holding it.

'It appeared to weigh between 300 and 500 pounds. I can only estimate the weight, as all contact was in waist-deep or deeper water. The animal did not have a "home base" and travelled throughout the area as long as it was offered some attention. It did not accept gifts of dead fish or exhibit penile erection. It did not tow boats or leap over them, but I expect it would tow a small boat if a rope was given to it.

'During my first experience with the animal, I went out into the bay accompanied by two of the village boys thumping on the side of a small wooden dugout. We then went into the water, hanging on to the side of

114

the boat and kicking our feet as the boys said this also attracted the dolphin. I explained to the lads that there were many large sharks in this area. They replied that since the dolphin had been near their village, no sharks had been caught anywhere near. Within a few minutes, I felt pressure on my leg and looking into the water saw this gentle animal. I put my arm around it. It raised its head out of the water and vocalised. It then began to swim close to us, allowing us to hold on to it. If we let go, the animal would return and remain close by so that we could continue physical contact. As more people from the village entered the water with members of my staff, the dolphin became even more responsive and began circling all of us and swimming amongst us. It allowed children to grab its tail and turn the animal completely upside down, holding its tail out of the water. It never offered resistance of any kind to anything the children did to it.

'I noted the dolphin continually vocalised and seemed to want to make serious contact. It appeared interested in having us go with it as it swam in larger circles into deeper water. It would surface, vocalise and wait for us to come closer. It appeared to want to show us something. It would tap you with its nose, rear its head out of the water and vocalise with varying degrees of excitement. It would always be watching for a response. The animal appeared to be bored quickly with human games such as ball throwing.

'I took several photographs of children with the dolphin and, in some cases, of children getting on the dolphin's back.'

In late June 1983, John Walsh returned to Costa Rica intending to record for analysis the curious vocalisations the dolphin was directing towards people. But tragedy had stricken. According to the 29 November 1983 *Weekly World News*, fisherman Raphael Conteras found the dolphin entangled in his net, calmly waiting to be released. Instead, he hacked it to death with a machete and took the carcase back to a horrified village. A week later, the story goes, sheltering from an electrical storm in the bottom of his boat, Conteras was struck dead by lightning.

Simo

As if retracing Donald's coastal odyssey, the next episode of a solitary dolphin in the United Kingdom took place near the tiny Welsh fishing port of Solva in Pembrokeshire.

Fishermen first became aware that there was a dolphin in the area in the spring of 1984. A young male bottlenose, it was called Simo, following the Greek tradition.

Solva is a popular summer resort and over the warm months of 1985 the dolphin's interactions with people intensified rapidly. Much of his behaviour follows patterns now familiar from previous episodes. He was curious about any underwater activity and would watch divers keenly. He would push canoes and tip people off airbeds. He swam in rapid circles around fishing boats and when parties came out to play with him he usually gave a welcoming leap. As people jumped in with him he would greet them by putting his head on their shoulders. At times he became most excited and would give them dorsal-fin tows, and belly-up pectoral-fin tows, wriggling when he wanted to be free. Sometimes he became so aroused he bit and drew blood, but if people behaved in a calm manner he responded accordingly.

His excitement was accompanied by sexual arousal. He would rub his erection against swimmers. If they became excited he would suddenly erupt into boisterous activity, butting them and even leaping right over them.*

In short, Simo was an especially approachable young dolphin and allowed people to touch him and frolic with him much sooner than dolphins of other episodes. A special behaviour was to swim belly up beneath a snorkeller with penis erect and lift him/her slightly from beneath. But it was clear that he adapted his behaviour to the capacities of the people he met.

Solva resident, Ann Marks, kept a diary of the dolphin's activities and her meeting with it during 1984.

A sample of her experiences recorded on 6 September 1984 (her third dive with Simo): 'I go in and Simo comes to greet me immediately, as he does now with anyone entering the water. I hold out my hands to him as he approaches. He allows me to stroke his head, beak and eyes — as I would a horse. Turning on his side, he presents his abdomen for me to stroke. As I tickle him gently under his pectoral fin, he actually flexes it. David and

---

*Leaping and butting of dolphins is a courtship gesture. We have an account of this towards humans, arising in a group situation. On 11 February 1986, Thelma Wilson had spent a happy half hour with a tribe of 30 bottlenose dolphins near Coromandel, New Zealand. She wrote: 'Then one dolphin started circling me and I was a little concerned. He seemed a bit larger than the others and was circling very close. Each time I left the surface he barrelled over the top of me, then leapt out of the water, coming down with tremendous force — easily enough to squash me if he miscued. On getting back into the boat Andrew commented that he thought I was about to be raped. The thought *had* crossed my mind . . .' (Project Interlock Newsletter No 38).

116

Simo play boisterously. Eventually Simo becomes quite rough. Rearing out of the water behind me, he smacks down almost on top of me, but doesn't touch — a great wave of water and a massive body much too close. Frightened, I head for the boat.'

On 30 October 1984, Simo met Clare Sendall, one of his favourites. After snuggling up to her shoulder and giving her a pectoral-fin tow, Simo dived deep, leapt out and came down almost on top of her. A glancing blow caused a headache. On 6 September Ann Marks spent an hour with Simo: 'He has started to blow massive bubbles underwater — some form of communication, I suspect. He only does it when close to us.'

On 18 September 1984, Anne and Eve Sendall met Simo:

'Eve goes in first and gets a great welcome. I follow. Simo goes down the anchor chain and standing on his head over the anchor, shows it to me. Then he whirls around to Eve. Rearing out of the water and wriggling with joy he returns to me to be tickled under his pectoral fins and, turning on his back, his abdomen. I stroke him from nose to tail. He returns to Eve, rests his head on her shoulder, turns on his back and gives her a pectoral-fin tow. Then he comes at me, head-on, opening and shutting his jaws, showing very neat, sharp teeth. I wonder what he wants and recall Horace Dobbs saying Donald would mouth things as though trying to understand them better. So I gave him my arm to mouth. Gently his beak closed around my forearm, holding it a moment before releasing it. I fondle his mouth and beak gently, talking quietly to him. We spend an hour together, until it gets too cold.'

On 26 September 1984, Simo was found playing, as he often did, with a resident seal and her pups, rolling and twisting among the rocks and kelp. 'As we drew closer to get good pictures, the mother seal got worried about her pups and wouldn't play. So Simo left them to encircle us and entice us further out to sea. Twenty metres out, he circled us again and played for a few minutes before returning to his seals. When we closed in for more pictures, he breached alongside, leapt high in the air and crashed down, covering us with spray. Then he led us out again, made a splashing between us and the seals and returned to his playmates. Point taken. We stayed put.'

Carola Hepp is a young German dolphin-enthusiast who came to visit us in New Zealand and then went on to dive with Percy, Simo, Jean-Louis in France, and Romeo in Italy, sending us accounts of her experiences. With Simo during 1985 she remarked on his adaptability:

'I saw him playing rough games with some American soldiers from the base nearby. Then Tricia Kirkman and I had a nine-year-old girl in the water with us. She was holding our arms for security. Simo circled us quietly. Spontaneously we began to chant. Initially it was unorganised but gradually we fell into a rhythmic, ululating wail rather like African women. All the while Simo circled us closely, listening as if bewitched. We felt a profound spiritual communion with the dolphin.

'The next time we met Simo, Tricia and I floated face down in our wetsuits, holding hands, her left in my right. We relaxed and meditated. Next moment Simo was there, touching my left leg with his beak, going up and down my lower leg with his mouth closed. After about five minutes, Tricia moved and said to me that her right hand was extremely ticklish as if somebody was giving her mild electric shocks. Now Tricia has a problem with diving — she always comes out of the water with icy, bloodless hands and has actually felt quite desperate about this. She even had electro-shock therapy to dilate the blood vessels in her feet. Well, as we came out of the water, she told me her hands were better — both of them. No longer did they feel extremely cold. We wondered what Simo had been doing?' (High frequency sound *can* generate heat — author.)

'A funny episode occurred when we were operating from a small boat in a mixed group. We had to urinate by getting into the water in our wetsuits. Cold water increases the frequency of urination, which makes things worse. I just could not pee with my trousers on. I cramped up, concentrating and suffering, while those in the boat laughed at me floating there. Simo swam in a curving arc, banged his body against me and suddenly I was peering through a huge brown cloud. Simo had defaecated directly at my body. I started laughing and peeing at the same time.'

Tricia had been induced by Percy to take up swimming when down in Cornwall. With Simo she developed a close relationship and over the summer of 1985, as Horace Dobbs and a film crew followed him, Simo led her on until she became a proficient snorkeller who spent every moment she could in the sea with the dolphin. Towards the end of the summer, Horace watched Tricia spend four hours in the water without a break; the dolphin never left her side for more than a few moments throughout the whole period. To Horace, the bond between the adolescent male dolphin and the mature female human developed into 'What I, as an objective, rational scientist hesitate to, but must honestly describe as, a non-sexual love affair.'

Tricia knew that if she was in the water other people would get no more than scant attention from Simo. She often climbed back into the boat so that others could enjoy his company. Everyone secretly hoped Simo would favour them more than anybody else. Even during a short encounter, Simo was able to generate the impression that he really did find each person special. Though their contact with him was brief, few visitors were ever disappointed.

From Kieran Mulvaney we received an account of an unusual episode with Simo that took place a short time before he withdrew from the area:

'There were a lot of people down to see the dolphin and they all got in at once. This meant there were really far too many in the water for one dolphin — about eight — especially as Simo seems happiest when he has just one person to play with. What made things worse this time was the presence of Tricia Kirkman. Simo adores this incredibly warm, sensitive woman, who can't swim the width of a pool, but clad in a wetsuit, gambols with Simo like a good swimmer.

'Well, from the beginning it was clear Simo wanted her to himself. When all the swimmers formed a circle Simo pushed Tricia away from the rest. He took her on about four tows within less than a quarter hour. Whenever anybody else tried to intervene he let them know they were not wanted. He bit two people, drawing blood, and he buffeted another around the head with his tail. Eventually everyone else left the water and he played with Tricia alone.'

Among the photos that Carola sent to us is one with Simo's tail protruding from the water alongside Tricia's head — a clear territorial gesture. At times Simo would try to prevent Tricia returning to the boat. It should be understood that both the bites and tail swipe would be intended communicatively; at full strength their power would be devastating.

Another picture shows Simo alongside Carola, who is floating on the surface in her wetsuit. Simo has an erection lodged firmly behind her knee — a new erogenous zone.

One game Simo shared with Sandy, the Bahaman, was that he would approach snorkellers on the surface from behind, rise up vertically in a headstand and hit them on the head with his beak, dislodging their mask and snorkel in the process, before sinking back into the sea. Wisely, those interested in Simo's welfare published a special pamphlet for visitors, warning them of Simo's vigorous antics and advising swimmers to wear buoyancy-aids in case he ducked them; to stay calm if he grew too excited, and to avoid hurting his delicate skin with any sharp, hard objects.

Towards the end of the summer of 1985, Anne Marks noticed Simo behaving strangely. He was sluggish and instead of chasing eagerly after boats he struggled just to come to the surface to breathe. Then one day, he surfaced slowly beside a boat, opened his mouth and sank just as slowly back into the water. She did not see him again. The locals all hoped fervently he had been pining for company and had rejoined a resident dolphin school a few miles up the coast. But some had a grim feeling they would not see Simo again.

The episode in which Simo monopolised Tricia Kirkman, driving off the others, interested me particularly. Those who sent accounts ascribe it simply to dolphin jealousy, but in the context of Simo's other interactions with people, it deserves a little more thought. We may assume dolphins are capable of communicating with each other in modes other than body language. Out of an ongoing context gestures can often be ambiguous. Scratch your ear in Italy and a man may take it you mean his wife is rendering him a cuckold at that very moment!

Autistic children have communication problems similar to a lone dolphin interacting with people. Out of sheer frustration, such children often resort to violent behaviour. I wonder whether Simo's behaviour was an attempt to communicate through body language, to all involved, that he preferred a one to one relationship? Meeting dolphins in their normal social groups, I have sometimes seen people behave as they would towards a film star or a famous person; they think only of themselves, become pushy and domineering, forgetting that communication is a two-way street. Lone dolphins would experience both aspects of human behaviour: the active and the receptive. After sessions of physical hijinks the latter may be more rewarding.

For Jan and me, the possibility of an interspecies ethical code or interlock etiquette has arisen. After many encounters with groups of dolphins, we find it better if no more than two people enter the water initially. If all on board leap in, it can be like a 'cocktail party' — just as a sensitive one to one relationship is developing, another person may dive down and cut in on the exchange. When dolphins approach us in a pair or trio, we have tried keeping the human side in balance; as more dolphins approach more humans join in, establishing separate groups of interaction. We noticed that when such 'manners' were observed, contacts lasted longer and the complexity of exchanges increased. Perhaps solitary dolphins deserve similar consideration.

Tammy

Just when the pattern of solitary dolphin/human interactions seems firmly centred on one species, the bottlenose, nature rings the changes. In New Zealand two other species come into the picture.

In 1981 at Whitianga, something odd began to develop when a female common dolphin, *Delphinus delphis* moved into the estuary and gave birth. Estuary-dwelling is without precedent for this offshore species. We will examine this still unfolding Dint later.

Then in autumn 1984, an equally unprecedented situation developed in New Zealand. Tamaki Estuary, in the heart of Auckland city, extends its arteries into suburbia. Dolphin lovers could scarcely believe their eyes when television news presented magnificent aerial displays of a dusky dolphin, *Lagenorhynchus obscurus*, at Tamaki.

Insulated with an especially thick layer of blubber, the dusky is a circumpolar species, an inhabitant of cool southern waters. Only rarely have small groups of around half a dozen been reported in the Hauraki Gulf, adjacent to Auckland. With a winter range normally extending only as far north as Hawke Bay, and a gregarious, open-sea lifestyle similar to the common dolphin, the dusky dolphin's solo residence in an Auckland estuary is almost as preposterous as a polar bear in the Sahara!

For any dolphin to survive a period of time in such close proximity to an urban population is equally incredible. All other Dints have been at reasonably remote locations.

It was around 24 March when Tammy arrived among the rows of moored pleasure craft at Tamaki. Water temperature was 65 degrees fahrenheit and seven weeks later, when the dolphin left, it had dropped to 58 degrees. At the outset, a fisheries officer claimed to have seen two dolphins in the area and there were other reports that Tammy arrived with a group of dolphins that left soon after.

During the initial few weeks the dolphin ranged over half a kilometre of the river but as time went by it seemed to concentrate around the bow of one particular moored launch — except when feeding. People began to approach the dolphin in dinghies, on surfboards and by swimming, but for the first weeks Tammy showed little response. He seemed to lay claim to a territory about four metres square at the bow of his favourite launch. Divers approached this area and those who showed special care would hover in the current a few metres from the edge of the dolphin's 'turf'. Tammy would approach them from behind and underneath, roll on his side at the surface to scrutinise them and then swim in a semi-circle. This would be repeated two or three times and then Tammy withdrew into his territory and the divers would enter.

It seemed like a greeting ritual on the threshold of his space, and those divers who were sensitive to Tammy's needs went on to develop close relations with him. Even at the outset, in early April, this ritual occurred at every approach and Tammy made his scrutiny of the diver even if no further interaction took place.

When people approached with less thought, putting pressure on his space, Tammy had an avoidance strategy; he disappeared into the murk and bobbed up alongside another moored launch further across the river. If surfboards or dinghies pursued him, he ducked under and returned to his original position. Visitors soon got the message and backed off. When a man in a dinghy approached unobtrusively and drifted about in the vicinity, not following or chasing Tammy, the dolphin showed him a lot of attention. People learnt quickly and the dolphin was able to establish the kind of relationship he preferred.

On the shore opposite Tammy's homebase live Linda and Craig Johnson, who became devoted to the dolphin and kept a log of his activities — the most detailed part being one period of eight consecutive April days. In the mornings and evenings and after feeding periods, Tammy's aerial behaviour was amazingly complex. No solitary dolphin has rivalled his acrobatic displays. He would make a sequence of superb arching leaps, then loop-the-loop underwater to erupt into the air in the opposite direction. On one occasion the Johnsons saw him perform 48 tail-stand leaps consecutively, his tail remaining in the water. They noticed that whenever Tammy made more than ten leaps he tended to bend his body in an 'S' shape against the sky.

A typical sequence from their log is: 'five sideways leaps combined with two spinning leaps (rotating around the long axis of the body, like the spinner dolphin), and finally a high double somersault, end for end, in which he rose to a height of five metres, measured against the mast of an adjacent boat. Delighted observers felt Tammy was proof that there is no need for dolphin jails.'

Clearly, the aerial displays had no part in feeding activities, which were quite distinct. In this they would have the support of Dr Bernt Würsig who has made a study of dusky dolphins in Patagonia and believes they have three categories of leaping behaviour: simple arching leaps in order to see whether seabirds were gathering on the surface to feed on baitfish; noisy splashing leaps to scare baitfish into a tight meatball for easy predation; and complex aerial displays that take place after feeding. The scientist felt that the latter was performed for pure joy, perhaps with a social function such as the bonding that comes from shared pleasure.

So, the lone Tammy out on the estuary performing before Auckland city ensured his own safety and the admiration of all who saw him, whether from the river banks, on nightly television news or in newspaper photographs.

It was noticed that, just prior to a major display, Tammy would first begin to rock on the surface alternately exposing his blowhole and slapping his tail — perhaps hyperventilating in preparation for strenuous activity. Following the rocking-horse manoeuvre there would be some surface splashing, either belly side up or right side up — then the leaps.

While feeding he often turned upside down and slapped the water with his tail, probably to startle mullet and sprats. He was seen with such fish in his jaws. He was also seen eating eels. During feeding sessions, he swam in circles, doing about three circuits slowly (45 kicks per minute) before moving to another area.

Dave Stephens is a computer engineer and diver who had often met dolphins in the open sea and would visit our home to discuss their behaviour. From the outset Dave dedicated himself to swimming with Tammy and attempting to interact with him. After his first five hours in the water with the dolphin, Dave came to think Tammy was indifferent to human company. Then, on

12 April, with visibility at only one metre, Dave extended his hand and the dolphin deliberately brushed against it for half his body length. For a week or two after that, Dave pursued body contact experimentally. Whenever he tried to touch Tammy in full view, the dolphin invariably withdrew to beyond his reach. On other occasions Dave tried to make contact when he thought Tammy was unable to see his action, as when the dolphin was rising directly under the diver. On no occasion did he, or any other person to his knowledge, ever manage to initiate contact with Tammy. Dave came to believe Tammy had a special capacity to sense the pressure waves of a moving limb.

Meanwhile the Johnsons observed him at play. He toyed with pieces of seaweed, a cardboard box and drifting logs. He balanced a leaf on his beak. Several times he responded to a tennis ball, pushing it with his beak, but when a girl, intending to toss a ball in front of him, accidentally hit Tammy, he showed obvious alarm and withdrew for ten minutes before resuming his usual two to three metres distance.

One day the Johnsons took a hydrophone out on the river to listen to Tammy underwater. From a distance, they heard a high-pitched scream like a seabird. As Tammy approached them the sound was lower in pitch like a strained cry. He then came close and squirted a solid jet of water into Linda's face with his blowhole.

On 21 April, Dave Stephens had his most exciting interaction with Tammy. He had been reflecting on the importance of mimicry as a form of communication. This day, before approaching Tammy, he prepared himself by practising the pre-leap behaviours until he could present a good rendition of the dolphin's rocking-horse/tail-slap manoeuvre, and the upside down splashing. At the top of the tide, with visibility at two metres, he swam out to Tammy's home base.

'As soon as I approached Tammy, I knew this time it would be special. Usually he investigates a swimmer and then goes about his business. But he stayed within ten metres of me and frequently swam a metre below, on his side, watching. I did my tail-slap, rocking-horse movement. Tammy surfaced about three metres behind me and watched, his eye above water. He then sounded, emerged a body's length away on my right and started tail-slapping in a semi-circle around me. Then he sounded again and began leaping and spinning. He did ten leaps between five and 50 metres from me and then quickly swam to midstream. Two friends on shore got photos of all this.'

Dave got a good look at Tammy this time. He was from 1.2 to 1.6 metres long (adult size is 1.8 metres), and clearly, he was a young male. Dave sent a drawing of his genital area — a slit about 100-150mm long and then a single anal aperture.

'Tammy is active at night and can be heard blowing out on the river,' he said. 'A man that fishes that area says eels, mullet and flounder are getting scarce in Tammy's range so we may soon see him move further afield.'

The next day Dave had his final swim with Tammy before he had to fly to Australia on business. He wrote:

'The tide was flowing in strongly, but I wanted to meet Tammy underwater this time, so I duck-dived to about two metres and dolphin-kicked up current. Tammy swam up behind, about half a metre away. I made no attempt to touch him. We just swam there, side by side until we both went up for air. This happened three times. It is hard to describe this meeting but I think you will understand. Then my friends arrived in the dinghy and we changed over. In all, we had a wonderful hour with Tammy on very close terms. He frequently had his eye out of the water, watching us. He also did something quite significant: twice he expelled solid jets of water from his blowhole. I thought this was strange at the time but later it hit me — God, he was mimicking our snorkels, just as I had mimicked his swimming.'

When Dave returned to Auckland, eager to resume, he learnt that Tammy had gone. Concerned, he spent a full day on the estuary in a dinghy, looking for any sign of the dolphin. Thankfully his search proved fruitless, confirming reports that Tammy had departed for the open sea.

Craig Johnson last saw him on the morning of 15 May.

Dave's final comment was: 'The fact that Tammy lived in a city of half a million people for nearly two months without being harmed must be a great encouragement for the future of human/dolphin relations.'

Certainly, and while Dave was away, dolphin lovers in Auckland contacted us in great alarm. A powerboat race was scheduled on Tamaki Estuary; a triangular course right through the dolphin's range. Neither city authorities nor fisheries were prepared to intervene and stop the race, in spite of the pleadings of dolphin welfare groups. Then Tammy's defenders did the obvious: they explained the problem to the powerboat people. Without a moment's hesitation, the club cancelled the event, not wishing to harm the dolphin in any way.

Jean-Louis

If Donald, Percy and Simo give an impression of how solitary male dolphins respond to humans, then Jean-Louis belongs with Opo and Nina. As this lone female dolphin interacts with humans over a period of 12 years, a gentle picture emerges (the best we have of this gender) which is in contrast to the more dominant and boisterous males. When we encounter dolphins in their normal social groups such gender differences are much more difficult to distinguish. For this reason alone, the story of Jean-Louis is invaluable.

Near the western-most tip of Brittany, in the region called Finisterre (meaning Land's End) is Baie des Trépassés, Bay of Lost Souls. Exposed to the full onslaught of Atlantic storms, its beaches receive the bodies of many seafarers who perish in these dangerous waters.

One spring day in 1976, Breton fisherman Yfic set out from Le Vorlène, a tiny, man-made haven in one corner of the bay, to work his lobster pots. On the way he thought he saw a shark fin, a Jean-Louis — the local name for the blue shark. Then, as he was lifting a pot, he felt resistance. A dolphin was teasing him by pulling on the rope. And so it began. Yfic named this 3.3-metre, scarfaced bottlenose, Jean-Louis, to the confusion of all who came to know her in the following years.

At first the local fishermen did not welcome competition from a dolphin, but gradually they came to reverse their attitude. For some reason, although bottom-dwelling fishes became scarce, they found more and more mullet around, which is strange, because mullet are a favourite food of dolphins. (Perhaps she discouraged other predators.)

For the Breton fishermen, Jean-Louis became a mascot and an omen of good luck. She accompanied them to sea not to steal their livelihood, but from sheer curiosity. As they went from pot to pot or checked their nets, the dolphin inspected their catches closely. (Nowadays she puts *crabs* in their pots for them.) Having won the hearts of the fishermen, Jean-Louis was relatively safe. Three strangers, who hunted her with a harpoon, had to flee the area like thieves. And if it is ever discovered who shot at her . . .

The first we in New Zealand knew of this dolphin was a letter from two London university students, Caroline Tombs and Andrew Crofts, who had read of our research. In July 1981, six members of their college canoe club returned from a holiday in Brittany with stories of a lone dolphin that had approached their canoe and surfed with them for several hours, frolicking around within a metre's range.

The following year Caroline and Andrew, with five other canoeists, set out for Brittany hoping they might see this dolphin again.

'We camped for a week at the Baie des Trépassés, a sandy beach enclosed by rocky coastline and deep, clear water. Often the bay is battered by thundering surf, but during our stay a large anticyclone persisted in the Atlantic and we were treated to clear skies, hot sun and calm seas. We abandoned surfing and took to sunbathing, snorkelling and paddling round the coastline.

'Every day lifeguards zoomed across the bay in their inflatable, and after a day or two, we noticed a lone dolphin leaping in its wake. This in itself was very exciting to those of us who had never seen a wild dolphin (Dauphin as we immediately christened it). Of course the others were even more excited, wondering if it was their surfing companion of the previous year.

'After Dauphin's second successive appearance in the wake of the inflatable, Andy, who had been closest to it last year, was convinced it was the *same* dolphin. Andy and I pulled on our wetsuits and got in our canoes — short surfboats with flat white bottoms. As usual the sea was flat as we paddled out of the bay towards a small, rocky harbour where some yachts were moored, and where we had seen divers in action. As we approached the harbour, I felt, rather than saw, a sleek grey fin lift out of the water, no more than two metres away from my canoe. I heard a great "whoosh" of expired air, and just as suddenly, the great body disappeared. My reactions of incredulity and fear were quickly dispelled by a jubilant Andy, who immediately started paddling furiously in the direction which Dauphin had taken. "She's playing with us ," he shouted.

'And indeed she was — an elaborate game of hide-and-seek which lasted an hour. Dauphin would swim slowly and deliberately, frequently appearing at the surface until we had approached within a few feet, then quickly dive and shoot at highspeed in another direction. Sometimes we were completely baffled by her disappearance, until we realised with a surge of exhilaration, that she was lying beneath us, just a few inches below the boat.

'Towards the end of that first session, we discovered a new game. We would capsize, hanging upside down in the water and Dauphin would swim up towards us, nearly touching noses, then swoop away. Eventually exhausted, we left Dauphin and paddled back to the shore. Dauphin followed us close inshore, then turned out to sea again.

'As we beached, the others, puzzled by our strange antics, were setting out to investigate. Andy and I forgot our exhaustion, flung our canoes, masks, snorkels and flippers into the van and drove round to the harbour. A despondent group greeted us. Dauphin had gone. Just as we were leaving however, Neil, vigilant on the cliff top, yelled, "She's here, she's here!"

'Soon all seven of us were in the water, playing the hide-and-seek game, with Dauphin rushing like mad from boat to boat, leaping out of the water in excitement. This time, Andy and I had brought our snorkelling gear; we put on our masks, fins and snorkels and flipped into the water for the best game of all. As we started finning and diving, Dauphin approached and dived with us, often within touching distance (although something held us back from actually touching). Sometimes Dauphin and I were diving in complete synchronisation, surfacing and diving together, as (we are certain) she deliberately slowed her pace to accommodate her clumsier companion.

'During this time, we were able to take a closer look at our friend. Dauphin was at least three metres long and seemed battered, with several large scars on the dorsal fin and head. I think that this second encounter was two hours long, with all of us exchanging snorkel gear to dive with Dauphin. When we tried the capsizing game wearing masks, we were able to see Dauphin approaching from below and almost touch the glass with her beak — fantastic! Again, sheer exhaustion tore us away from our playmate.

'Two days later, our trip was over. We left Brittany after another day's "interlocking". Surfing-wise, the trip was a disaster but we were agreed that we had had one of the most wonderful experiences of our lives. We are convinced

125

that the dolphin deliberately approached and consciously played with us in an attempt to communicate. We were also aware of some emotional interplay, I could not call it telepathy. I can only describe it as a vague feeling of emotional upheaval, and great tenderness.

'At the time, all these things seemed too strange to be true, and we could hardly believe them except in the group. Of course, we have since found out that our experience was by no means unusual in the field of dolphin encounters!

'Reading your book *Dolphin Dolphin* I was fascinated to hear of so many meetings similar to our own, and it was enthralling to see our own experience in the light of the whole picture of cetacean encounters. I am convinced that interspecies communication between humans and dolphins *is* happening. We are inspired; we are charged with new ideas and are enthusiastic about our return to Brittany and more dolphining.'

In April the following year (1983), three of the young canoeists made a return journey to meet the dolphin. This time they learnt that their Dauphin had been dubbed Jean-Louis by the locals.

'The pattern of encounters was slightly different from the year before, with just three of us — Waldo, Andy and myself. We usually approached from the harbour, one person paddling a canoe, the other two snorkelling. Jean-Louis seemed to "hang-out" by the large rock on one side of the bay — Dolphin Rock. She would appear there soon after our arrival in the vicinity. Often, it seemed as if we could "call" her by bashing a paddle on the deck of the canoe; we also tried blowing whistles and calling but this didn't seem to work as well as the canoe technique. We all got the feeling that she knew who we were, partly because of the growing intimacy of our contacts. She came much closer to us in the water and stayed close for several minutes, whereas last year she seemed to zoom up for a quick look then zoom off again. She also did some spectacular leaps and twists in the air, often as we were leaving, almost as if she were saying goodbye. One day she actually leaped over Andy's head — wow!

'Conditions were *very* different this time; the sea was rough and the swell made it difficult for us. I'll never forget the moment when I first saw Jean-Louis again. We only half believed that she'd still be there, and were preparing rather lethargically for our first dip when we saw her out in the bay. I immediately experienced a tremendous feeling of excitement/exhilaration/elation, yelled out something like "whoopee" and dashed into the water. She seems to have such a capacity for generating joy in the people she comes into contact with. We noticed this in all the people we spoke to — fishermen, divers, local people — she's everyone's sweetheart.'

By the high summer of 1983, word of Jean-Louis was out. Divers from all over Europe were arriving. In July, the remote bay was packed for its short summer season. So many film crews were present that to avoid an international incident, the French and British had to come to an agreement — British in the mornings, French in the afternoons. On some occasions, Jean-Louis would be surrounded by as many as 20 French, Dutch and German divers. From dawn to dusk her life was filled with humans.

When Andrew, Caroline and Waldo returned in September of that year, they were shocked to find Jean-Louis quite changed. She would buzz up to them and swim away with little of the responsiveness she had shown back in April. They wondered if the summer and her growing fame had been too stressful for her.

Among the divers who came to spend most time with Jean-Louis is Francois Pelletier. A professional film-maker and photographer, Francois has travelled widely to document marine mammals and their relationships with man — from the symbioses of the Imragen people and dolphins in Mauritanea, to that of river dolphins in the Ganges, Indus and Brahmaputra, and the seals of Newfoundland.

Francois found that his initial contacts with Jean-Louis were brief, but eventually she began to stay little more than a flipper's length from him. With so much publicity, as time went by, he found she became both more trusting and more suspicious. Francois was quick to appreciate that touch had to be on her terms entirely. When she was feeling sexy she would glide between his legs and rub against his wetsuit. She would put her beak up to his mask, withdraw and shake her head — a gesture he came to recognise as an invitation to play. Then she would lead him on a chase, extending his powers to the utmost, pausing, head above water, to encourage him, before resuming the game. In poor visibility she would charge towards him at high speed and at the last moment, leap vertically from the water to land a few metres from his fins.

A terrible tease, she almost drowned Francois with laughter when he saw a scuba diver searching everywhere for the dolphin while Jean-Louis was directly behind him, her beak almost touching his tank. When Francois and his companion were caught in a strong surface current and had to make a bottom journey back to the boat, Jean-Louis guided them, constantly passing to and fro between them and their goal.

Early one chill autumn morning, Gerard Soury and his wife, Clo, climbed down the steel ladder on the jetty at Le Vorlène wearing full scuba rig. On the way they rapped on the rungs to attract Jean-Louis. Down below, Gerard was wrestling to secure a loose scuba tank when he found himself face to beak with the dolphin. Seconds later, as Clo was hunting for her gloves, the same happened to her. The divers swam out through the kelp forest to a plain of sparkling white sand at 15 metres, with the mooring chains of fishing craft rising here and there.

Nearby, Jean-Louis nose dived into the sand, did a quick turn and approached Clo, again beak to mask. Wanting to make eye contact, Clo withdrew slightly. But Jean-Louis advanced. To Gerard it looked like a tango; the more his wife receded, the more the dolphin approached. Then Clo made a major error. She tried to *stroke* Jean-Louis. With a wave of her tail the dolphin disappeared. Five minutes of 'deprivation punishment' elapsed before she rejoined them beside a mooring chain.

The innkeeper at the local 'Wreck Robber' had told them that Jean-Louis would accept no gift from a diver, not even a fish, but one thing she really adored was music. Down on the sand Jean-Louis had seized a mooring chain and was shaking it with an undulating motion. The links made a musical clinking sound. Vertical, her beak in the sand, Jean-Louis seemed to sleep. It was ecstacy. From time to time, an eye would open to show she was still alert. Then she would give the chain another nudge with her beak.

As a final touch during the divers' ascent, Jean-Louis nose dived at the sand, rose like an arrow and burst through the surface out of sight to fall back in a splash of white water. Suddenly she stopped, right between them, as if to say, 'Go ahead — take your photos. I won't move!' Then she escorted them back to the jetty ladder with a continuous to and fro-ing.

In July 1983, a major dolphin expedition set sail from Cornwall to Brittanny. James Wharram took his 18-metre Polynesian catamaran *Te Hini* over to act as a stable platform and base for a television team. The plan was to film Dr Horace Dobbs performing a whole series of experiments that would explore the dolphin's capacities.

Horace wrote to us: 'After a few days we realised there was no way we could direct Jean-Louis' attention to what we wanted her to do — in contrast to Donald. So in the end the film director, Peter Gilbe, would say to me, "Go in the water Horace, and see what she wants today." ' In the credits of their fascinating film, *A Closer Encounter*, Jean-Louis appears as assistant director!

Horace saw that this dolphin had a personality quite different from Donald's. In his view she was totally feminine: 'Not only did she get her own way,' he felt, 'but she was much more subtle than Donald in achieving it. I tried hard to get close to Jean-Louis but she would not allow more than fleeting physical contact, even when we were playing a complex game.'

The film shows Horace with an ingenious underwater xylophone made from a row of inverted wine bottles, partially filled to varying levels with scuba exhaust, floating from a base line and anchor. The idea was that Jean-Louis, following his demonstration, would pick up another small bottle swaying on a tether, and use it to strike musical notes. Jean-Louis showed not the least inclination to involve herself with this contrivance, nor any of the other elaborate game devices Horace assembled. She snubbed their invitation to aquaplane, a sport Donald loved. But she showed real interest when Horace clicked two sea shells together and when he juggled a sea urchin on a knife blade. Then she signalled with body language that she wanted him to follow and led him on the same sort of chase Pelletier experienced. And while the film crew were reloading cameras, Horace saw her appear out of the haze with a plastic bag draped on her pectoral fin. She would release it and then alternately pick it up on her beak and fins as she circled. Gradually Horace began to feel, in contrast to the aquarium situation, that *he was the performer*.

A fine sequence shows the dolphin leading him through the kelp jungle around her home-base rock, and through the cascades of white water in a complex game of follow the leader/hide-and-seek, with her creeping up on the diver unawares. At the outset, he had grasped her body language gesture to play. Approaching the diver's head, Jean-Louis gently rolls on her side.

Capturing such episodes on film can be immensely difficult. For this reason *A Closer Encounter* could be an invaluable resource for future researchers in this field — much like the frame-by-frame analysis made by anthropologists, Mead and Bateson, of old film in which Balinese elders teach children traditional dances.

Among the international crew aboard *Te Hini* was our German friend, Carola Hepp, who sent us her account:

'In the mornings my first action was to look around for Jean-Louis. She always came after a while. Perhaps she would be out near her favourite rock where the waves always foamed. That is her jacuzzi, her spa pool where she is caressed by millions of tiny bubbles. Sometimes she tries to lure selected people out there and then plays games with them. Around midday, for two hours, she often withdraws, probably to her rock. Some days she disappears in the afternoon. At night I would hear her breathing outside the hull in which I slept. One night I was filmed as I sang my dolphin song on deck.

Finally the film director was satisfied. I stopped singing and strumming my guitar. At that moment Jean-Louis leapt out of the water high above the illuminated deck.

'Jean-Louis likes to caress anchor ropes with her genital region; she glides up and down the taut line with her eyes half closed. If she ever thrusts her head out of the water it seems to be in the course of a game. She seldom accepts touch and actively avoids an outstretched hand, but at times she would lie right on the surface, totally still, listening or looking, as I lay alongside her. Sometimes in such moments I could touch her.

'One time she didn't mind how much she was touched. For underwater sequences, our film crew used a video camera with a cable running up to the recorder aboard the ship. The camera motor produced certain frequencies and Jean-Louis pressed her beak on the port. The cameraman had trouble getting her away so he could film. He shoved her aside, gently kicked her in the sides and eventually grabbed her beak and pulled her away — all things we would never have dared. But she would *not* be distracted. She showed interest in the transmission cable too, moving along it with her beak. It would have been conveying frequencies she could sense. Eventually she seemed to appreciate that we needed her in the pictures and co-operated.

'Sometimes, if she were not there, we would call her by rattling a chain. Approaching, she would wriggle her head from side to side, listening to the sound. When it stopped she would hover motionless, eyes closed, her body stiff and tense, as if enraptured. When I tried clapping two large barnacle shells together, the new sound held her interest. With her eyes wide open, looking up and down at me, she held her body still, close to my head.

'One day I sat on the rocks watching her play with a snorkeller. She would lead him straight towards a wooden boat. Following her, he banged his head on the hull. This was repeated several times, with him diving to avoid a collision. Afterwards both snorkeller and I compared notes and were agreed that it seemed she wanted him to dive underwater — so he did.

'With the scientific experiments, Jean-Louis refused to react as expected. Perhaps there was a lack of spontaneity and she instantly became bored when all those elaborate preparations led to nothing exciting, on her terms. Is the scientific approach too tight, leaving too few possibilities? Life is not just "yes" or "no". There is a truth somewhere in between what we know but often refuse to acknowledge because it can't be proven. Science perhaps is too restrictive for a creature that has a different kind of intelligence. I now feel we should have played more with *her*, not with devices but spontaneously with our bodies and all our sensuality and art, with music and dances full of joy. Then perhaps she could have shown us her capacity to dance, her games, her body and what she knows. Who is more intelligent than a creature that seeks to teach those who just want to teach?

'My theory about spontaneity was reinforced at the very end. One day we had to refilm the arrival of expedition ship *Te Hini* in the bay, this time with Horace Dobbs and the film crew aboard. As we set out to do this, Jean-Louis followed, but with no great fuss. I think she knew we were returning. But the day *Te Hini* finally departed for England, Jean-Louis followed a long way out, leaping high above the water before she disappeared in the wake.'

At the same time as the British film crew were present, but for a much longer period, film-maker Francois Pelletier and his wife, Catherine, were also filming Jean-Louis — in this case a dual approach. While they were documenting her as a solitary marine mammal, they also made a fantasy film involving a beautiful young girl, Onde, swimming like a dolphin and wearing a large monofin.

One day, a young Frenchman came to our home in New Zealand, a Greenpeace sailor by the name of Franck Charreire. He had been diving assistant to the Pelletiers during their film-making. Having quite a long acquaintance with Jean-Louis, Franck was a great help in preparing this account. He told us Jean-Louis did not mouth people's limbs like the male dolphins in this history. She would thrust her beak into armpits and knees and between the legs from behind, but such physical contact only occurred after he had been diving with her every day for two weeks. Then she would allow him to stroke her side, accelerating past but returning to repeat the performance. With Onde, their young actress, Jean-Louis swam in close unison, while they both moved in the dolphin manner and surfaced to breathe together.

Rather like Nudgy and Donald, Jean-Louis would swim at high speed behind an inflatable, maintaining a position just beneath the propeller. Then she would accelerate and leap in front of the boat.

Each summer, since her popularity in 1983, people flock to the Bay of Lost Souls to enjoy a romp with Jean-Louis, or to watch her antics from the shore. Typical would be this account by 14-year-old Emma Jones:

'My swim with a wild dolphin has to be the most exhilarating and wonderful experience of my life. On the last day of our holiday in Brittany we arrived at the tiny fishing haven. I tied Dad's spanner and wheel-brace around my waist and carried my flippers and mask, a little embarrassed, down the steep rocks and waited. There was Jean-Louis, escorting a fishing boat into port. Quickly I adjusted my gear and jumped into the cold Atlantic. I trod water and clanged the spanner and wheel-brace together, making a lovely din underwater. Suddenly the crowd on the jetty were shouting and pointing in the direction of a blue and white fishing boat. I swam over, still clanging. As I neared I heard strange whistling and clicking noises. Suddenly, out of the green gloom she came. I was face to face with a three-metre dolphin, so close I could have touched her. She was engrossed in the sounds I was making, floating vertically, her eyes half closed like somebody appreciating music.'

In August 1985, Dutch diver Louis Robberecht made his third visit in successive years to Jean-Louis. This time he decided to attract her attention with an odd musical contrivance. He bowed the serrated edge of his diving knife over a length of copper tube, like a 'singing saw', alternating it with taps of the flat blade. Jean-Louis approached within centimetres, head down in a vertical posture. To Louis' surprise, this time she did not withdraw when he tried to touch her, as she had done previously. Wondering whether movement would hold her interest as well as sound, he began sweeping his collection net in circles, fast and slow, just in front of her beak. For over an hour he held her interest with a constant variety of actions and sounds, including human speech. Louis concluded that Jean-Louis would interact as long as she met with novel behaviour.

In October 1986, our friend Franck Charreire made a special journey to Brittany in chill autumn weather to provide us with an update on the Jean-Louis situation. He and a friend met the dolphin out of the bay in 10-15 metres of water with four to five metres visibility. For a while they circled, all three diving and breathing in unison. Jean-Louis remained close to them, eye to eye, sometimes within 50 centimetres. Franck was enchanted by her tranquil, fearless gaze. Often just before surfacing, she would release a large bubble from her blowhole and the two divers tried to copy her gesture.* Whenever they stopped, she would do so too, her body upright in front of them. Franck was astonished at her quiet, languid manner. It seemed she had grown accustomed to divers and was much less flighty since his last visit.

Suddenly she disappeared, to return with a plastic bag wedged on her pectoral fin. She stopped in front of them and released it. Franck took it and they began swimming, followed by Jean-Louis, her beak almost touching the plastic. The divers would exchange it with each other or just push it in front of them. Jean-Louis was most excited. She swam vigorously defaecating and releasing a bubble from her blowhole at the same time. She never took the plastic bag back, just kept nudging it with her beak. Again the trio swam in unison before the session came to an end, with the arrival of other divers.

Franck spent that afternoon observing her interactions with scuba divers and snorkellers, as Jean-Louis shuttled to and fro among them. He noticed that with divers who did not know how to interact, she would inspect briefly before going to play by herself with a mooring chain. On the hill top above, a happy crowd was being adequately entertained. Whatever the dolphin did pleased them and they cheered and clapped with delight each time she surfaced to breathe.

During a solo session two days later, Jean-Louis led Franck out towards her rock. As he followed through the green haze she returned, a plastic bag held to her side by her pectoral fin: 'She gave it to me like a present. I accepted and began the same game as before, swimming in unison with her, our eyes only 20cm apart, and holding the bag out just in front of her beak. Then I dived down eight metres and released the bag with a little air in it. Jean-Louis made a spiralling ascent, following the bag.'

Franck repeated the game several times until, to his horror, he noticed his camera was filling with water. In his excitement he had exceeded its operating depth. In conclusion Franck remarked that it was curious he had never heard Jean-Louis make any whistles or echo-location clicks during any of his visits:

'Perhaps, because of her lonesome life, she has lost the habit of vocal communication, but it mystifies me as to why she would not echo-locate in those murky Atlantic waters?'

1987: Jean-Louis Revisited — Since 1983, when he last saw Jean-Louis, Andrew Crofts longed to return. But in the meantime he visited us in New Zealand and we discussed the significance of certain dolphin gestures such as the bubble-gulp** and I showed him Franck Charreire's latest account. Andrew felt he needed the right people to share the experience with — sensitive, open-minded, games-players. Eventually he set off with three friends, including Kitty, who had met the dolphin in 1982, for a brief five-day meeting with Jean-Louis.

*Franck notes that the bubble-gulp occurred when she was excited after synchronous swimming or some such joint activity; defaecation seemed to be an invitation to follow; she would do it in front of you and vanish.

**These solitary dolphin episodes don't always follow each other in time, but overlap. Thus, when Andrew visited us in late 1986, I explained what we had learnt from the *Rampal* interactions, wishing that our communicative insights could be tested with Jean-Louis.

'On 12 September 1987, I found myself paddling out through huge surf with a fresh southwesterly wind, hoping to meet up with Jean-Louis. The tiny surf ski was really unsuited to the conditions, but used in the hope that she would recognise it. As I rounded Dolphin Rock, a massive set of waves came in, making my position precarious, the more so without a friendly dolphin there to reassure me. Suddenly I heard a "whoosh" and there she was three metres away, and then, further away. Not wanting to lose her interest, I slipped on my mask and capsized. She swam up out of the gloom and sussed me out. Then she came closer and let out a rush of air from her blowhole, which I had not seen her do before. After a while she would come very close, gazing intently, bubble-gulping occasionally as she surfaced. After an hour, in which I had been getting to know Jean-Louis again, Kitty and Suzanne joined me in snorkelling gear. Meanwhile, Jean-Louis had been sharing her attentions around with other people in the bay. She soon accepted the new arrivals by bubble-gulping, defaecating and making noises.

'A game developed between Kitty and Jean-Louis with a mooring. Jean-Louis would go to the bottom of the main chain and slowly rise to the surface. If it was tugged she would then let out a massive bubble-gulp, obviously enjoying the whole experience. When I swapped places with Kitty, I tried rattling the two chains together. She came up looking delighted, squirming with pleasure and bubble-gulping. All through this encounter she had been getting more and more excited. Then, taking me by surprise, she jumped halfway out of the water behind me. She disappeared for a while and suddenly shot up to me at high speed, dropping off a plastic bag from one of her pectoral fins, in an obvious invitation to play. I picked it up and swam over to the mooring where we had been before. She was there with another plastic bag on her left pectoral fin. This was getting interesting, but was interrupted by shouts. I looked up to see Suzanne nearly run down by a boat and shouting that she had cramp in her leg. So the game was abandoned. We left the water soon after.

'My immediate impressions were that she was more confident than before and more excitable, although she still avoids physical contact. Through the bubble-gulp she was giving a visual signal to us, the meaning of which, and the way we should respond, were unclear. Games like the plastic bag routine, seemed the best way of developing our relationship with her.

'The next day I went in first with Marie, in snorkelling gear. The chain mooring had been lifted so I tried tugging on another one. Jean-Louis came up and touched the rope about six metres down. I tugged it and she rubbed against it, just behind her blowhole. As I kept tugging she turned over, rubbing her chin, pectorals and most of the front of her body on it. She worked her way up towards me until we were nearly touching. At times she had her eyes shut and certainly seemed to be enjoying the experience.

'We then went on to swap snorkelling gear with Kitty and Suzanne, as arranged. They continued the game. Jean-Louis did some impressive tail slaps when she seemed to be excited. Kitty went off for a swim with her, alongside and behind her at times. Then Jean-Louis slapped her tail and shot off, returning very fast, with her mouth open, oscillating her head and body in an up and down movement. Kitty was disturbed and frightened by this and decided to get out. As she made her way in, Jean-Louis came up slowly and bubble-gulped once, which on reflection, seemed to be a conciliatory gesture.

'I went back in and had a fast swim with Jean-Louis; with our bodies close, we surfaced together to breathe. She would lead me back to the rope mooring for a rub as before. If I headed off she would follow. I tried rolling on to my back. She mimicked me, rolling over, then flicked her tail and shot half out of the water. Quite a display of her power, certainly one which I could not follow. She was, I felt, showing us who was the boss in the water. I left, exhausted.

'On the 14th, the game changed again due to Jean-Louis' insistence. The weather had improved by now and the sea had calmed down. She appeared with a ten-metre fishing boat and was totally engrossed in descending and ascending its twin mooring ropes, twisting around them at times. She showed no interest in the rubbing mooring of the previous day, but joined me for some synchronised swimming. Kitty developed quite a rapport, blowing bubbles with her at the mooring. Jean-Louis responded by sometimes letting out a trickle of bubbles rather than the usual "gulp". I wanted some photos of her but was frustrated by the poor visibility at the surface, so I used the rope to climb down. She joined me and would ascend with me when I ran out of air.

'To finish off, I decided to paddle the surf ski back to the beach. To the dismay of eight divers who had just arrived, she followed me out, frolicking and bubble-gulping below me. When we got to the corner, where the waves peaked and crashed on the rocks, she gave me an impressive display of her surfing skills.

'On our last day, after a protracted mutual bubble-blowing session, Jean-Louis nuzzled Kitty's arm with her beak, one of the few touches during any of our encounters.'*

Kitty recalls: 'I tried singing "Greensleeves", clicking, growling, talking, "ooing", humming. Of course, all these involve bubbling. What I most wanted to tell Jean-Louis was how much I felt for her and how lovely she was and how pleased I was about her spending time with us. The noise that came closest to that was a sort of soft purring; a low, fuzzy "pooo", bubbled gently out of my lips in a kiss. It was close to the sort of noise my cat loves and I was inspired by my conversations with him. It is the noise that babies love. Anyhow, Jean-Louis was fascinated. Whether it was the noise, the tiny controlled bubbling, the emotion or the vibration, I just don't know. When I did it, Jean-Louis liked to come close and "rest" by me with her beak pointed to my face. She sometimes seemed to *slowly* roll over. I did it hanging onto a rope from a boat, pulling myself up to get more air before going down again to carry on bubbling.

'Because I was looking at her face and sometimes had my eyes shut (like you do when you kiss), I didn't notice that Jean-Louis was letting out streams of tiny bubbles like mine, until Andy pointed it out. While I was bubbling at her, I was promising I wouldn't touch her, as well as telling her I loved her — with *thoughts*, not words. It was after a long session of this as Jean-Louis rested floating closer and closer as we went up and down with the waves that she finally decided to let herself touch me with her beak.'

Andrew resumes: 'Later she showed no interest in either the fishing boat mooring or the rubbing mooring, but was entranced by a new game. This was started by a man who had caught our attention the day before, when he nervously wandered about, wearing only swimming trunks, talking to himself and preparing a small unseaworthy inflatable boat. He had Jean-Louis in

*Kitty had been trying to make noises underwater that would interest Jean-Louis.

raptures when he dropped a tiny grapnel anchor on a long rope into the water. She would faithfully follow it down and then back up, bubble-gulping, only occasionally spending time with us or any of the other divers who were about. It was a busy day at the "petit-port", with many visitors, even Television Francais, who were filming from the shore. Eventually we had to go to get the ferry back to England. How could we say goodbye?

'After 11 years in the same small area, Jean-Louis is thriving. She seems to be handling her popularity better now than four years ago, when I felt that she showed the pressure of a busy summer. She dictated the games that were played. I tried many times after the first day to follow the approach that she had made, but she showed no further interest. In her use of the bubble-gulp I think she shows that she realises humans are visual creatures. To communicate with her through sound as well would be exciting.

'I noticed her defaecating gesture most when she left me after synchronised swimming sessions and got up her speed of travel again. Kitty thought it was a sign of intimacy and sharing, "letting everything go" as she sensuously rubbed on the mooring ropes while we pulled them.'

Final words on this still unfolding story go to Gerard Soury as he reflects on his meeting with Jean-Louis: 'This corner of the world is privileged to have an animal that has left its social group to live apart. Jean-Louis has established a territory and, it seems, doesn't wish to leave. She has set up the most unusual relationship with man. She alone determines the game rules. These same humans, elsewhere thirsty for power, now submit to her and even share the privilege with their own kind. How can this creature achieve this when lofty ideals and human values have not succeeded for any length of time? She has spontaneously brought humans together with no goal other than the sharing of beauty.'

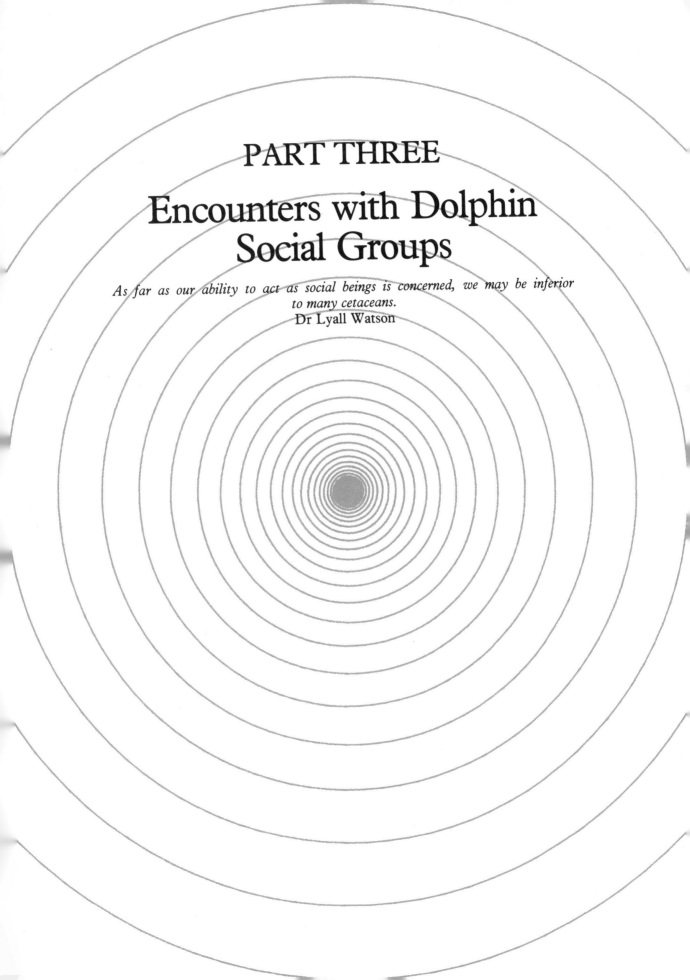

PART THREE

Encounters with Dolphin Social Groups

As far as our ability to act as social beings is concerned, we may be inferior to many cetaceans.
Dr Lyall Watson

The Monkey Mia Dolphins:
A Family Affair

On the opposite side of the globe from the Bahamas, about midway along the strangely bleak western coastline of Australia, there is a vast bay, swarming with marine life, named Shark Bay by English buccaneer William Dampier. It was here that Dutchman, Dirk Hartog, landed on 25 October 1616, and nailed a pewter plate to a post: the northern tip of Dirk Hartog's Island, Cape Inscription.

Peron Peninsula thrusts 130 kilometres into Shark Bay. On its sheltered coast, at the end of 160 kilometres of rough road, in the middle of nowhere, is Monkey Mia caravan park. ('Mia' is Aboriginal for 'home'.)

And there for more than two decades the most unusual of all human/dolphin friendships has been quietly unfolding. Were such events to occur anywhere other than in this remote corner of the globe, publicity and human pressure would probably have overwhelmed the dolphins long ago. The rugged Australian outback took care of that until recently.

At Monkey Mia, even as you read, wild dolphins (*Tursiops aduncus*) are visiting the beach by the caravan park, mothers and babies in knee-deep water accepting fish with an audible squeal and transforming people of all ages into glowing children. This is the longest Dint on our records and these dolphins break much of the Dint pattern. They accept gifts of dead fish; they maintain a social grouping; they accept human contact, although for many years they refused to allow swimmers to touch them, and seemed to avoid people wearing diving gear.

Ben Cropp — One of the first to write about the Monkey Mia dolphins, was underwater explorer, Ben Cropp, in 1978. Cruising off the Western Australian coast in *Beva* in a cold 30-knot south-westerly, he sought the shelter of Monkey Mia.

As he approached the anchorage there were bottlenose dolphins on his bow but when he slowed to anchor they left and sped straight for the beach. Right up in the shallows, they were begging for feed from several fishermen who were cleaning their catch. One fisherman waded in, handed a fish to a waiting dolphin and gave him a scratch on the back. The dolphin stayed there as if wanting more. Five dolphins lingered in the shallows, fins slicing the water among the boats at the edge of the beach. With his camera and some fresh fish, Ben and his friend, Lyn Patterson, approached.

Lyn waded out with a couple of fish. The dolphins swept straight in and one nudged her leg as she dropped a fish into its waiting jaws. Then Lyn slipped her hand under the dolphin's jaws and kissed its forehead. It exhaled a blast of spray in her face. She gave it the fish in her other hand which it swallowed and then circled back to her. This time Lyn held the fish in her teeth and let the dolphin gently pluck it from her mouth.

As a former owner of a dolphinarium, Ben could scarcely believe his eyes: 'With captive dolphins it takes up to six months of intensive training and lots of kindness and attention to achieve what Lyn had done. But the Monkey Mia dolphins are in the wild with the open ocean right behind them. Yet they choose to meet and trust the people who come here.'

Ian Briggs — The next person we heard from was Interlock friend, Ian Briggs, who visited the area that same year (1978): 'At present,' Ian wrote, 'the resident colony consists of nine adults and three young. They hang around in the Monkey Mia vicinity throughout the year except from November to February, the hottest, windiest period, when they leave, presumably for breeding purposes.* Even during this period they make a day visit every couple of weeks.'

As far as Ian could tell, the exact origin of this human/dolphin exchange is a little uncertain as it goes back some years and involves people who were not in contact with each other. It seems a young girl on holiday began to feed some dolphins around the fishing jetty many years ago. (Accounts vary from 1957 to 1972 but centre on 1964.) Eventually she lured the dolphins into the beach and regularly fed them. One dolphin became extremely friendly. 'Old Charlie' the fishermen called him, although he may have been a female. This dolphin became a legend in Shark Bay and would let children sit on him as their parents snapped photos.

Charlie would turn up at the end of the 150-metre jetty at precisely 7.15am and round up the bony herring, nosing them towards the fishermen. Keeping them closely packed while the men jag-fished, he would only pick off the injured herring that broke loose. If by chance he himself were jagged he would come in and almost beach himself while the hook was removed. Then at 8.30am, practically on the dot, Old Charlie would scatter the herring in all directions and that would be the end of bait fishing for the day. Nobody thought anything of it — it was just the accepted thing that Charlie rounded up the bony herring each morning for everybody's bait.

Then, as with so many legendary dolphins, Old Charlie was found dead on the beach. Some say he had been shot, perhaps by a holiday-maker resenting his presence near a net, but his death *could* have been from natural causes.

Since then other dolphins continued to visit the beach.

Hazel and Wilf Mason — In 1976, Hazel and Wilfred Mason became proprietors of the Park and from them Project Interlock has received a series of letters and tape recordings, even a visit from their son, Noel. They first wrote to us on 1 March 1979:

'We moved to this area from Perth in 1976 to take over a run-down small caravan park in a sheltered bay. We arrived in January and occasionally a dolphin and baby came in close to the beach. We were amazed to find we could feed it but were soon advised by the locals that this was quite normal — ever since "Old Charlie" had been tamed. We started to piece the story together. There are so many people who now say *they* taught the dolphins to come in and feed. Personally, we think it would take longer than one short holiday to achieve this. Possibly it was the joint effort of quite a few before they became as tame as they are at the moment.

*This was later disproven.

138

'Our first year progressed and we found that the dolphin and baby ("Beautiful" and "BB"), came in more and more as we kept feeding the mother. At this stage the baby was never allowed closer than 15 metres to the beach but one day Hazel was down there swimming when the mother brought her baby in. For three-quarters of an hour, Hazel had the pleasure of swimming up and down the beach with the mother, who kept the baby near to her.

'After that, the baby seemed to be allowed to come in much closer so, perhaps, Hazel was being introduced.

'As that first season progressed there were more and more dolphins coming in but mainly mothers with babies (which makes us wonder if Charlie might have been a mother who had lost her baby). At one stage that year, we were feeding 11. We only supply a small portion of their daily requirements and would never consider more because they must not become dependent on us for food.

'For the people who stay at the caravan park it is wonderful because everyone can feed them by going down and standing in half a metre of water. They come right in, as our photos show, and open their mouths, receiving a pat or scratch while being fed. People are amazed that such large animals come in from the wild and feed so gently from their hands. They become more demonstrative when the holiday period brings larger audiences. They are the biggest "hams", the more people on shore, the bigger the show. If cameras are pointed at them they stick their heads out, often four dolphins in a line.

'One day some children came up from the beach to tell Wilf that Speckledy Belly, the oldest in the group, had a hook in her mouth. She had a big snapper hook jagged in the side of her mouth and. a long wire trace hanging out. I felt I had to do something so I grabbed a pair of pliers and called Speckledy Belly in. She came right in on the sand and lay there, quiet and trusting while I struggled to remove the hook. It took me a good 30 seconds twisting with the pliers and she never moved. I had the feeling she knew exactly what I was doing. When the hook came out she shook her head, swallowed a couple of fish I gave her, and moved out, wriggling backwards into deeper water.'

In 1977 a new dolphin swam up to Hazel making an unusual sound:

'She was definitely crying. I tried to feed her but she wouldn't hold the fish in her mouth. I tried several ways but she could not swallow anything. A few days later we found her dead on the beach. Her jaw had been broken. Unable to feed, there was no way the dolphin could have survived.

'After three years the babies, who were small when we arrived, are now weaned and taking fish from us. It will be interesting to see if they return in this coming year or whether they will go their own way, perhaps, to return when they have babies of their own in later years. We wonder if this is a "dolphin" nursery area. Will the mothers who have now weaned their babies withdraw and other mothers with young babies arrive? (The answer to this was to be forthcoming . . .) I hope this gives you some idea of the set up here. The best of luck with your project. Regards, Wilf and Hazel Mason.'

Among the beach-visiting dolphins when the Masons came to Monkey Mia were the three mothers: Beautiful, Holey Fin and Crooked Fin, with their respective offspring — BB, Nicky and Puck. For the initial two years there was another, much older dolphin, with hardly any teeth, old Speckledy Belly — sex unknown.

Beautiful would bring her nursing baby BB to meet the humans. In their company, moving slowly among their legs, she would still nurse her baby.

Holey Fin was the most trusting and friendly of the dolphins. It was Holey that Lyn Patterson held and kissed at first meeting. (Holey has a hole in her dorsal fin.) Her baby Nicky, like Beautiful's BB, was born around the time of the Masons' arrival. The calves: BB, December 1975; Nicky, December 1975; and Puck, December 1976.

The Masons were extremely concerned about getting special legislation passed so these dolphins could be protected. A full-time custodian was needed to keep an eye on the public, teaching them how to treat the dolphins, what to feed them and what not to throw to them. Many dolphin deaths in captivity are caused by foreign objects being thrown to them. Speed boats needed to be kept out of the crucial area and all netting banned. Project Interlock supported their drive for special protection.

Publicity began to focus on the Monkey Mia dolphins through television and print media, postcards and travel brochures. Word was out; people began to make pilgrimages from all over the world to meet the friendly beach dolphins of Western Australia.

Chi-uh Gawain — We next received a letter from a globe-travelling yoga teacher, Chi-uh Gawain of San Francisco. Chi-uh had read Ian's article on the dolphins in *Simply Living* magazine and had made the long journey to Monkey Mia to spend a week with them. While there, she read our correspondence with the Masons, and kindly offered us her experiences. (Subsequently, Chi-uh visited us at Ngunguru, came to sea on RV *Interlock* and made many journeys to Monkey Mia — so deeply can dolphins affect people.)*

She wrote:

'It is wonderful to be here and watch the dolphins turning smoothly just offshore. They whistle through their blowholes to attract attention. When anyone wades into the water they come up to greet him with heads just out of the water and open their mouths if a fish is offered. They will accept it from anyone. They seldom eat whiting, mullet or pieces of fish, preferring bony herring.

'If they don't want the fish they swim away and come past two or three times within touching distance. If you try to touch them they often swim quickly out of reach, but sometimes will glide slowly under your hand to be stroked — especially if they know you or have just accepted a fish. *I don't think they take them because they are hungry.*

'They are in good condition and always look full. People say, "What can I do with the dolphins?" and all they can think of is to offer them a fish. They don't always eat the fish but treat this feeding as a game, sometimes going off and dropping the fish underwater. I have only once seen a dolphin actually swallow a fish that I offered but Hazel had selected one they specially like.

'They are especially attentive to children. Provided they stand in water up to their knees people can touch and pat them.

'They come in every day, sometimes seven dolphins, but always at least three. Usually in the morning — sometimes before dawn but at other times, not until 10am. They leave in the afternoon but have often been seen near the jetty at night. If the weather blows up they usually stay clear. Fishermen

*Her book *The Dolphins' Gift* (Whatever Pubs, Mill Valley, 1981).

clean their nets in the morning and feed the dolphins any fish they don't require — this may account for the morning visits.

'They did seem to hang around a little closer the day I stood waist deep and whistled the same little tune for about 20 minutes — I'm *not* a good whistler, but when I tried it again today they were not interested.

'When I first came they didn't know me at all, but five came to greet me, all accepting fish, and two let me touch them. One came back several times and glided under my hand, as though a cat having its back scratched.

'But when I go swimming among them, they ignore me or at the most, make fast circles around me at three to five metres distance. Sometimes they come close to Hazel and Wilf, and cavort around them, but nobody, not even the Mason family, can touch them while swimming and *they won't tolerate anybody near them using face masks.*

'I have the impression the dolphins don't want us to have any *purpose* in our relations with them. And especially, no tricks or pretence. (Scientists please note.) They want us to be ourselves and they themselves, and to meet as friends.

Om Shanti, Chi-uh Gawain.'

In her book, *The Dolphins' Gift*, Chi-uh explains how she came to believe the Monkey Mia dolphins did not want contact with swimmers. Following that initial beach session with her, the dolphins withdrew to deeper water. She decided it was time to try a swimming approach. "I had this fantasy of swimming right in among them and having them jostle me, as I had seen them do to each other.'

But that did not happen. Although the dolphins eyed her, she did not feel welcome. As she swam towards one it would dive and disappear. She turned to another, but it withdrew. Then the first would begin circling her rapidly within two metres range. When it took off, another swam rapidly towards her, dived and surfaced on the other side. On a speedy pass they sometimes came quite close. 'But always,' she wrote, 'they stayed at least, beyond my fingertips.'

On her second day at Monkey Mia she again tried swimming. The dolphins circled her — 'but out of reach'. After three days on the beach, Chi-uh had been able to touch three mothers and two juveniles, but *never* while swimming.

From our many encounters with dolphins we now realise that this avoidance of touch is quite a normal response towards swimmers. Once trust has been established dolphins may accept contact but strictly on their terms and usually on their initiative. For many people acceptance of touch encourages efforts to grasp and ride. Such dominant behaviour would seem inappropriate early in a relationship. Encountering humans on the beach, it seems the dolphins had learnt they could control the situation with people remaining stationary, while they were mobile.

In Cornwall, Percy (1982-4) had learnt to encounter swimmers from the outset and he avoided the beach situation. The day that Opo (New Zealand 1955-6) first allowed a woman she knew well to hold her in shallow water, three men got the idea of lifting the dolphin ashore and carrying her away.

1980 — When the dolphins returned to Monkey Mia for the 1980 season, the Masons were able to answer some of their previous questions. Not only did Holey Fin bring in her two-week-old baby, but also her previous offspring, four-year-old Nicky, now weaned.

'Normally they keep their babies well out but she let it come right in between our legs. I (Hazel) was talking to a chap when "Little Joy", as we have named the new babe, swam to and fro between us! We were swimming in chest-deep water — eight people and five dolphins cavorting together.

'Yesterday Wilf and I went in with face masks. Crooked Fin and her baby were around but would not stay while we wore masks. So far this season (8th March), we've had five here most days — Holey Fin, Little Joy and Nicky; Crooked Fin and Puck. There's a new adult dolphin Snubnose, who occasionally comes in for a fish. (Later found to be a male.)

'Twice now we have had the experience of telling the dolphin who was not eating a particular fish to "bring it back and change it" and the dolphin picked it up and dropped it at our feet. After four years of talking to them every time we feed them, we wonder if they understand — as a dog would. They certainly make noises as if trying to talk to us when feeding.'

Letter from Chi-uh — In August 1980, Chi-uh returned. For 12 days she kept a detailed journal. Towards the end she wrote to us:
'It is such a treat to be here again, to walk out to the beach, wade into the water and to be greeted by two or three dolphins swimming up to my knees and lifting their heads out of the water to be petted; and watching them frolic with one another.

'They are spending much more time here every day than they did when I was here for a week 16 months ago. I don't know if that's just seasonal, or whether they are becoming more and more friendly as the years go by. This time I'm privileged to spend 12 days here, and am keeping a detailed journal of my experiences with the dolphins.

'There are 12 dolphins that I can call by name and recognise by their appearance and personalities, and as many as 15 were counted here, all at once, on a day about two weeks before I came. Seven dolphins are real pets, and come regularly to take fish from anyone's hand and allow themselves to be touched. Joy, the new baby now eight months old, is here every day with her mother, but Holey Fin doesn't allow it to be touched. She nudges it away if it comes close to shore.

'The dolphins keep away from swimmers and if anyone puts on a face mask, they leave. I don't know if build-up of trust will ever change this.

'It is really marvellous to be so affectionately greeted when I stand knee deep in the water. They will even rub against my knee, and let me stroke all along their sides, swim away and come back for more — all this *daily* from dolphins who then swim off to the open sea, and who have never been captive or trained. Love Chi-uh Gawain.'

Chi-uh's Journal (August 1980) — Illustrated with a superb range of monochrome photographs, Chi-uh's journal from *The Dolphins' Gift* (pages 63–173) provides the basis for subsequent studies of the dolphin families at Monkey Mia. Her descriptions of variations in maternal care and offspring development, continued over several years were of particular interest to scientists and would eventually lead to more formal observations.

In 1979 Nicky, aged three and Puck, two, kept close to their mothers, but by 1980, Nicky was independent. Puck still kept closer to her mother than either of the other two juveniles at the same age. In 1979 BB, aged three, appeared to have no particular relationship with his mother and by the following year he visited the beach much less frequently than she did.

In 1980, the three juveniles — BB, Nicky and Puck — were spending a lot of time together, especially the latter pair, and Chi-uh observed several of their play routines.

In the game 'dolphin roll-over' the trio would swim fast abreast. The middle one, often Nicky, would then roll over one of the others, turning upside down in the process. For 'sideways leapfrog', Nicky and Puck swam side by side. Then one rose head first and slid over the other obliquely.

Some games could be adapted to include humans. Puck was making passes in front of Chi-uh, raising her beak above the surface to have her 'chin' scratched. Each pass increased in speed until, as Chi-uh reached beneath the beak, the young dolphin used her hand as a fulcrum to make a somersault, landing on her back with a splash.

Acceptance of fish gifts from tourists would often elaborate into a play routine, as when Nicky received a fish from a boy, severed its head and brought the body to Chi-uh, laying it in her hand. Offered back it was swallowed.

When the juveniles became too boisterous either mother would discipline them, slapping the water with her tail and clacking her jaws. It seemed Holey Fin and Crooked Fin often shared 'child care' responsibilities and their offspring had an especially close relationship. If newly born Joy came close to human hands, the mothers and the other juveniles would push her away. She related closely to them all.

In their transactions with people, the dolphins expressed themselves with sound and body language. When mouthing Chi-uh's fingers, Holey Fin made a soft 'oo-oo-oo' on a descending scale. Nicky would approach with a rapidly stuttering soft squeal, in response to Chi-uh's clicking noises. If children's fingers got too close to a dolphin's blowhole, it would splash its tail sideways or give a big puff, jerking its head up and veering off only to return again to the shallows — lesson established.

Dolphin mothers usually kept very young offspring away from human contact and one child who touched Joy had his fingers lightly nipped — which he accepted in good heart.

On a busy Sunday, when the dolphins met a forest of legs on the beach, they avoided contact, moving excitedly among the crowd, taking fish and dropping them and showing irritation at efforts to stroke them. But they *could* have gone away. During tranquil periods, Chi-uh would find them gently mouthing her fingers, wrist or toes. On one occasion she neglected to make her usual communicative noises, kissing and chucking sounds, and found none of the dolphins would let her stroke them.

Publication of Chi-uh's book in March 1981 provided an excellent reference for future visitors. From that time, an increasing flow of people made the long and arduous journey to meet the dolphins at the desert edge of the Indian Ocean. To our delight many of the keenest ones broke their journey across the Pacific at our place.

Letter from the Masons (5 November 1980) — 'We have reached the stage with the dolphin language (noises) where we know when they are happy, irritable or upset, just by the tone. For example, they make a certain sound when we want to feed them a fish they don't want. They let it slide out of their mouths, give it a prod with their beaks and always make the same noise.

'Little Nick has been exceptionally friendly lately. Often, after feeding she will come in, mouth your fingers, raise her head from the water and plonk it on your hand for you to tickle under her neck. Then slowly she swims through your hands to be rubbed along both sides. She circles and it is repeated.

'An added attraction here at the moment is a pair of cormorants, 'shags' we call them, who are so friendly that when you are trying to feed the dolphins they snatch the fish from the dolphin's mouth. Many a tourist has lost a fish, intended for the dolphins, to them. The other day I saw a chap knee deep in the sea, stooping to offer a fish to the dolphins when Heckle (could they be anything *but* Heckle and Jeckle?) dived between his legs, grabbed the fish and took off. He got such a fright he nearly did a back somersault.

'Recently, I saw a young dolphin become agitated when a diver was in the water. The mother, a hundred metres out, sped in and kept between the diver and her baby until he left the water.

'Not long ago, I met people who run a sheep station 60 kilometres south of here. They had set a net overnight to catch mullet. On checking in the morning they found two dolphins enmeshed but still alive. They had to be cut out of the net. The first dolphin released, circled the men in waist deep water and returned to the trapped dolphin, lying against it until released. Then they swam away together. Not knowing our dolphins, the people could not say whether they were from here. We must circulate fin markings among all the people in this region.'

1981 — When Chi-uh Gawain returned to Monkey Mia for six weeks in September 1981, she made an important discovery. Until then it was possible to determine gender only with the mother dolphins. She wrote to us in excitement:

'There have been several new developments to share with you. One item: I learned that five-and-a-half year old Nicky, and four-and-a-half year old Puck are both females. We had all wondered about the gender of these two young dolphins, one the offspring of Holey Fin, the other of Crooked Fin. Both had been coming to the beach at Monkey Mia almost every day of their lives, brought there by their mothers from their earliest babyhood. I even went so far as to buy a little Minolta Weathermatic to try to get underwater snapshots of their underbellies to try to see if we could determine their sex. (They won't let anyone wearing a mask get close to them.) Well, what had never occurred to me was to ask them to turn over and let me see!

'They do occasionally roll over on their backs in play when they are quite close, but they are usually moving fast. So, one day during the first week I was back at Monkey Mia, I was stroking Nicky as he/she swam past me over and over, and I said aloud, "Nicky, would you turn over and let me see whether you are male or female?" Meanwhile I was picturing in my mind the dolphin doing this. Immediately, there was a dolphin right in front of me, belly up, gliding very slowly past, and I could clearly see the two teat slits on either side of the anal/genital slit! All we have to do is make clear what we want. Then Puck also turned over for me, and she is female too.

'Wilf and Hazel have been terribly busy this year. Usually, they have a quiet time between their May and August school holiday-time. But this year the caravan park stayed full, and it wasn't until the end of September that they began to get some relief from pressure. The paving of the road is making Shark Bay much more accessible.

'The Dolphin Welfare Foundation is determined to get a permanent warden on the beach by next season. I had a good meeting with their executive committee.'

PLATE 17A   *Credit: Debra Glasgow*

For more than three decades at a caravan park on the West Australian coast bottlenose dolphins, *Tursiops aduncus*, have been approaching the beach and accepting gifts of dead fish.

PLATE 17B   *Credit: Debra Glasgow*

Three of the Monkey Mia females are *Puck, Nicky* and *Holeyfin*, her mother (from top to bottom).

PLATE 18A

Credit: Debra Glasgo

Holly, especially demonstrative and tender towards people, comes to the beach with her first fish proudly displayed in her beak.

PLATE 18B

Credit: Debra Glas

Nicky, born in December 1975, has spent her whole life visiting people on the beach. In November 1987 she gave birth to *Nipper*, and brought her baby in to the beach a few days later.

Credit: Debra Glasgow

ithout limbs *Nicky* is still able to express pride and maternal affection towards *Nipper*, her new born.

PLATE 20A

Credit: Debra Glas

For people like the beach warden, Sharon Gosper, who have developed an in-depth acquaintance, fish gifts are not need *Nicky* and *Puck* engage her in the 'seaweed game', a reciprocal exchange which several dolphin species extend to humans.

PLATE 20B

Credit: Debra Glas

Nicky plays with a morsel of eel grass. Recently dolphins have been observed balancing sponges on their beaks for prolong periods, a behaviour that puzzles scientific observers.

In the 11 months since her previous visit Chi-uh found the mother/calf relationship had not changed much, but whereas Puck and Nicky used to romp together or swim side by side, their relationship had altered markedly. Nicky tended to bully Puck, and her mother Crooked Fin would quickly intervene and lead Puck away. Joy, now one-and-a-half years old, still appeared to suckle. While approaching people in the company of other dolphins, Joy would not accept fish and rarely allowed human contact.

Chi-uh noticed that Beautiful was pregnant again; her son BB, five-and-a-half, seldom visited the beach. Nicky was spending a lot of time with Beautiful, who was not her mother. It seemed that contact with humans was changing Nicky's manner. In comparison with the others, she made three times more contact with the tourists, received twice as much fish and actually swallowed more. Nicky performed more acrobatics, bit people more and could be quite demanding and aggressive at times. With selected people she could be equally demonstrative and gentle.

Chi-uh saw the dolphins play the seaweed game, bringing a scrap of weed to a person, which if returned would be carried around on pectoral or dorsal fin, or in the mouth before being returned to the person. (For the dolphins' safety the Masons actively discourage the introduction of man-made toys, beach balls and the like.)

Jim Hudnall and Ann Spurgeon — Late in October 1981, we received a visit from American cetacean researchers, Jim Hudnall and Ann Spurgeon, who were returning from Monkey Mia after some interesting experiences. I had dived with Jim in Hawaii where he introduced me to the spinner dolphins of Lanai. We shared the same views about the need for a communication model in our approach to cetaceans, and Jim has had extraordinary success in filming whales and dolphins. We found we had all been taught the same lessons.

Prior to their visit, I had discussed with Jim the puzzling refusal of the Monkey Mia dolphins to accept swimmers and divers, and we speculated that it was probably the result of a human failure to communicate acceptable intentions. We knew from our open-sea encounters with four dolphin species that a direct approach, and attempts to touch, were most unlikely to succeed — it had to be a gradual and sensitive introduction as with any other animals, including human children. We have strict rules about touch with our own species, and different cultures have their own greeting rituals. Yet, somehow, it is hard for people to extrapolate such etiquette to a creature that looks so like a fish. With cetaceans we tend to have a block, a human-centred blindspot.

Jan and I sat down with Ann and Jim to discuss their visit. I was most interested in discovering how they overcame the 'mask barrier.'

In his dry, deliberate manner, Jim told us: 'Well Wade, I find it hard to believe there really *was* any mask barrier. When we got there we were eager to film the dolphins underwater but didn't want to disturb relationships already in progress, so we just watched for several days as people waded out and stood there with dolphins coming in to them. We had been told the dolphins didn't like people to swim with them and that they would leave the area if anyone approached wearing mask and snorkel.'

When the tourist season closed and things quietened down, Jim put on his wetsuit and floated in the shallows in a crouched position. The dolphins were with Ann and some other people, but they came over and gave Jim a sonar scrutiny. Day by day he familiarised them with his camera and, eventually, his mask and snorkel. They chose light-coloured masks and introduced them gradually, first wearing them on the forehead while still on the beach. They avoided wearing sunglasses so that eye contact was maintained. Jim did not give the dolphins any fish and Nicky would sometimes come up and nudge his hand before seeking fish from another. She was curious about the sound of his 16mm camera, which he ran empty for her benefit.

'Well, finally I put on all my gear and swam offshore with my camera. Visibility was about three metres. After a few minutes, suddenly there was a dolphin swimming in tight circles around me. I spun around with it. As soon as I got in synch, it reversed. When I matched it, again it began looping in the third dimension, diving under and coming over me. I was foxed; I could not perform on the vertical plane because of my camera. So that was it, just like any meeting with dolphins but much more friendly and closer than those in Hawaii — more like those I had dived with in the Bahamas.'

Jim felt sure that acceptance was due to familiarity with him and his gear in the shallows where the dolphins had ample opportunity to echolocate him. Meeting later in deeper water, there was no problem.

But one day, they saw four snorkel divers out among the dolphins. Each time a diver saw one, he headed directly at it, arm outstretched. A scene of pursuit and avoidance evolved with the dolphins trying to approach people on the shore but unable to get past the snorkellers.

Ann had not brought her wetsuit, thinking the water would be much warmer: 'The first time I sat down in the shallows, after standing for several minutes with the dolphins around my legs, they moved off quickly, *but as I made no attempt to follow*, they cautiously returned. Ultimately, I was able to lie down among them and snorkel out. It was like swimming in a fog. Occasional shadowy forms, then a dolphin face would appear next to me. Nicky! She always circled me first, then we would swim and breathe in unison. Once I was closely circled by Holey Fin and Joy. The baby was trying to nurse, swimming just beneath and slightly to the side of its mother.'

Ann noticed that as they neared the beach Joy would detach and wait offshore while Holey Fin interacted with the tourists.

As time passed Jim was fully accepted as a swimming companion. One day he went out to film and took a fish along for Nicky. But he came back to shore with two fish! When he had offered the fish, Nicky vanished, only to reappear shortly after with a fish for him. Her behaviour indicated that he was to keep both fish so he graciously accepted her gift.

As they were exploring the fish transaction ritual on the beach, Jim and Ann noticed and recorded certain sound patterns the dolphins were producing with their blowholes, in air. Two short, rising whistles seemed to inquire: 'Do you have fish?' and a single tone was emitted on acceptance. When a mother wanted to warn people of unacceptable behaviour such as patting a hand near the blowhole like a dog, or grabbing a dorsal fin, one of the few points where a human could restrain a dolphin, she produced a loud buzzing, snapping her jaws and giving a light nip.

Gradually they noticed individual dolphins had food preferences. Nicky would accept unthawed fish which the others had rejected. Perhaps, they surmised, growing up as a beach visitor, Nicky had been obliged to accept what older dolphins abandoned, and had developed an odd taste. Nicky liked a yellow tail species but hated one with a blue stripe.

One day Ann met her when she had three of a species the dolphin detested. After being offered them one by one, Nicky was clearly disturbed but kept nuzzling her so Ann laid all three fish on the bottom in a row and said 'sorry'. At this, Nicky made eye to eye contact and emitted an excited sequence of ten different squeaks and whistles, as in a conversation, lightly nipping her hand.

Jim began to suspect that fish acceptance was really just a game, since they often abandoned the gift unless it was a preferred species. Prior to their arrival, a dolphin had brought in a species much prized by humans for food, but it hadn't been accepted. On another occasion one was given to a delighted young girl who took it home for tea.

Jim concluded that when a human offered fish it was a signal to the dolphin of an intention to interact in the customary Monkey Mia way. Acceptance by the dolphins indicated a willingness to participate. People often came who offered no fish but they still enjoyed frequent contact. Jim felt that, since there was a constant flow of 'untrained' humans to the area, such symbol manipulation would be desirable. Human behaviour could not otherwise be predicted through habit. To Jim, the dolphins' gift of fish as a human food suggested 'a level of comprehension by the dolphin on which I will leave the reader to speculate.'*

One rare windless day Ann and Jim paddled their inflatable kayak some four kilometres offshore. They wanted to solve another problem — where was the main pod from which the beach visitors came. They were watching a group of dugongs munching sea grass along the banks of a channel when they heard a distant splashing; a pod of dolphins moved rapidly with high leaps. Curving towards the kayak, it split into two groups and one stopped, milling around in their vicinity. Among them, they saw Beautiful, who had not visited the beach for some time. With the other group was Notch, another beach visitor. This scotched the popular theory that the inshore dolphins were social outcasts.

Towards the end of their three-week study they saw some fishermen beach a 12-metre fishing boat which belonged on a mooring near the beach. Jim was impressed that for half an hour three dolphins watched the men closely as they worked on the hull. They were not interacting with the dolphins who seemed held there by sheer curiosity.

Ann's conclusion on the Monkey Mians: 'We looked often into the dolphins' eyes and the quality of the look they returned was unlike that of any animal we have known. . . It is possible that, if the human species can become more gentle, open and compassionate towards other life forms, Monkey Mia could be a prototype for the future — a voluntary exchange, mutually rewarded and inspirational.'**

In December 1981, Goldie was born to Beautiful.

1982 — In March 1982, Chi-uh Gawain again wrote to us from Monkey Mia: 'Wilf Mason wants me to tell you about Beautiful's new baby. She brought it in to see them on 10 December. It still had crease marks around it from being folded in the womb, and was golden in colour. Beautiful made no attempt to stop it touching Wilf's hand. They sent a birth announcement to the

---

*'Interpersonal Contacts With Free Cetaceans, Pt. II Human-Dolphin Contacts, Monkey Mia and the Possibilities'. Paper submitted by James Hudnall, Whales Alive Conference, Boston 1983.

**'Some of My Best Friends are Dolphins' — Ann Spurgeon, *Whale Watcher* Journal, Fall 1981.

newspaper naming it Goldie. Within a few days it became a normal grey, but the name persists. Marine biologist, Dr Ken Norris told me that the golden colour occurs sometimes in very new-borns, and the baby couldn't have been more than a few days old, perhaps only a few hours.

'Well, I had to come over from San Francisco and see it. I was quite sure that Beautiful was pregnant when I was here in October, but some said, "Oh, she's always been fat." I'm delighted she was, indeed, pregnant. The two come nearly every day, and the baby is not turned away even when it comes within centimetres of us. Six-year-old Nicky spends much time with them, and plays with the baby even more than the mother — a charming relationship.

'The baby is surprisingly long already, at three-and-a-half months old. We undertook to measure them, at least approximately, by holding a string as they swam slowly past, and tying knots at the points we estimated were head and tail — not very accurate, but a good approximation. Beautiful is about 214cm or about seven feet long, while the baby is 117cm, or just under four feet long; more than half its mother's length. We made a guess at measuring some of the others, too, and found that Holey Fin is 223cm, or about seven feet four inches. Her six-year-old Nicky is about 185cm or just over six feet; and her two-year-old Joy is about 138cm, or around five-and-a-half feet long. Not very accurate, but it gives us some notion of growth. It *seems* to us that Goldie is a bigger baby than Joy was at that age, but hard to tell. If so, is this one a male and Joy a female?

'This week we have been coaxing the dolphins to allow us to swim with them again (not diving, just ordinary swimming) by carrying fish in our hands and occasionally giving. Much love, Chi-uh.'

During her stay Chi-uh noticed that Goldie, the new baby, spent quite a lot of time drifting and frolicking alone, in contrast to Joy who had usually kept close company with an adult. Puck, five, and Crooked Fin were still surprisingly close, with Nicky still biting and bumping Puck whenever they were together and Crooked Fin intervening. Beautiful's six-year-old son BB seemed to have left the nursery group and only once visited the beach briefly along with a large group of offshore dolphins.

Rachel Smolker — In November, American dolphin researcher Rachel Smolker visited us after her first three-month study at Monkey Mia. Rachel had been working with Richard Connor, another graduate student, under the supervision of Dr Ken Norris of the University of California. Equipped with binoculars, cameras, cassette recorders and hydrophones, they began the first formal study of the Monkey Mia dolphins' social system and it would be continued for successive years.

Rachel told us that at first she found the situation rather stereotyped and boring, wondering how on earth she would endure three months of it. But after two weeks she found herself intrigued at the subtleties of dolphin relationships and personalities, and their varying responses to people. 'It was like a huge soap opera,' she said.

In summation, almost nothing could be predicted; there were not really any regular patterns. For example, the dolphins would at times be most responsive to an aggressive, domineering person and quite negative to somebody gentle. Nicky was especially moody. At times she seemed to work herself into an affectionate mood, accepting caresses, nibbling gently and nudging at genitals, most often those of human males. Usually, this was when she was the only beach visitor. At other times Nicky would be the most aggressive, domineering and greedy of dolphins, nipping people until she drew blood; rushing in from six metres away to intercept a fish intended for another dolphin.

One day Nicky befriended a man whom Rachel felt was not particularly agreeable. After lavishing attention on him, she went out and returned with a fish gift. She bit off the head with its dangerous spines, and gave him the best part.

Rachel found Puck especially responsive to swimmers. While snorkelling, she would circle ever faster and closer so that all Rachel could do was spin, facing her. Then she let Rachel place a hand on her flank, pushing back so that she assisted the human in spinning rapidly. Rachel noticed that when strange dolphins came in from the offshore group, Puck and Nicky would interact more intensively, as though showing how many fish and how much touching they could accept from humans.

The researchers had two unusual experiences in trying to determine the gender of certain dolphins. On one occasion Rachel found two adults (about whose gender she was curious) circling with their undersides towards her — unmistakably males. Richard became keen to learn, for certain, the gender of Puck. The dolphin tilted her belly to his face at 12 centimetres range — a female.

They found that at night, even in moonlight, the dolphins would not interact as they did during the day, even though in the close vicinity.

One day Rachel was sailing a surf catamaran when it capsized and the mast stuck in the mud. Two of the dolphins, Nicky and another, dived down and next moment the mast was free. It was most likely that they had nudged it loose. She was told the story of a 13-metre yacht stranded out on the tidal flats for three days. When it floated free, two dolphins came and guided it out through a maze of channels, one diving repeatedly by the bow, acting as a depth indicator, the other tapping either side of the bow to indicate direction. Not knowing where to go, the humans chose to accept the dolphins' signals.

Wilf Mason told her how he had set a fish net and the dolphins had herded a mass of bony herring into it until it was loaded.

Towards the end of her stay, Rachel noticed that at times there would be many people on the beach but no dolphins for days, so she could not attribute beach visiting to the presence of people. Some days there would be 150 people and just one dolphin — Nicky.

During 1982 we received visits from non-scientists, such as Edith Howland, from San Francisco, who had travelled widely — from the Bahamas to Hawaii — to meet dolphins in the wild. From Monkey Mia she wrote to us: 'The thing I feel most astounded by right now is the way they are a natural part of the landscape here. When you look out across the water you so frequently see their fins emerging and slipping back. Then, when to their presence is added their apparent interest in contact, the place begins to seem on the one hand magical, and on the other, as it should be. To stand in the water with a dolphin right there is so overwhelming and so simple all at once.'

Then Edith told a story against herself. As she stood in the sea facing Nicky, her mind full of mystical thoughts about dolphins, in a moment of pure Zen, Nicky nipped her finger, earthing her instantly.

And Georgia Tanner from Los Angeles: 'On my last day I had a chance to be alone with them and they did not open their mouths for fish as they do for most people. They seemed to know my ideas were different. They circled close to me as if offering pure love and emotion. When they left I wept. Then they came back again . . .'

1983 — By 1983 the pace of development was gathering at Shark Bay. The Masons and the Dolphin Welfare Foundation fought battles to prevent extensive development and beach mining.

In January our friend Pat Hindley, Professor of Communication Studies at Vancouver, arrived there. After meeting Nicky and watching her frolic with a shag, she wrote to us of the local politics surrounding the dolphins and the growing need for protection.

'Monkey Mia is quite astounding — as ecologically intact as I have seen anywhere. It is a special place on this coast, not only for its dolphins but for the richness of sea life generally — loggerhead turtles, large schools of manta rays, dugongs, sharks, including the fairly rare lemon shark, and huge molluscs known locally as "bailers". I'm helping Wilf track down the smaller ones so he can tag them. Then there are stromatolites — the oldest and biggest display of living fossils anywhere in the world; a 110-kilometre stretch of tiny shells beach; crazy tides that puzzle oceanographers; and special kinds of sea grass which are being studied.

'I'm going to talk to some people who run the post office who have been involved in shark studies here to see what behavioural data they have. Anyway, it seems clear to me that Wilf and Hazel are on the right track when they talk about the need for a National Marine Park. A mining company is seeking access to 125km of Shark Bay's high and intertidal shoreline, which will ruin the place.

'Listening to Wilf talk about the rapid increase in tourists into Denham, plans for a new black-top road, the increase in pressures on the caravan park and the expected numbers of day trippers from Denham, one becomes even more convinced of the need for international protection.

'Watching people with dolphins gives me some concern, too — despite information available around the place, there's a lot of ignorance and often the desire to photograph little Johnny feeding a fish to a dolphin over-rides the rules of common sense.'

In August Chi-uh Gawain arrived.
'So here I am for my fifth visit. They are coming every day, as usual, staying for varying periods of time. Last Thursday there were eight all at once. On Saturday there were three that stayed from 7am to 1pm, while three little girls were swimming much of the time back and forth along the shore. Two of the older dolphins, Holey Fin and Snubnose, would loiter alongside them only about a metre or two away.

'Beautiful and her baby Goldie, who would be about one-and-a-half years old now, don't show up anymore. I feel pretty sure that Holey Fin is pregnant again. If she does have another calf around Christmas time,★ that will be her third to be raised at this beach. It's lovely to see how the family histories unfold over the years.'

During her stay, Chi-uh was able to identify 19 dolphins: the four adults and five juveniles that 'come to hand' on the beach, and a further four adults and six juveniles that came in but remained aloof. Good companions once

★Holly was born in December 1983.

150

more, Nicky and Puck seemed to have resolved their differences. The male BB, now seven-and-a half, once again was making beach visits, accepting hand strokes and the occasional fish, even though his mother, Beautiful, no longer visited.

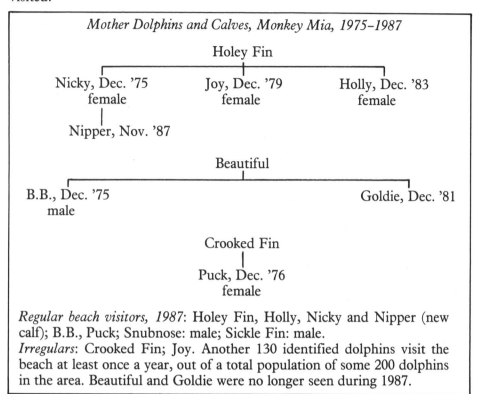

Mother Dolphins and Calves, Monkey Mia, 1975–1987

Holey Fin

Nicky, Dec. '75 female

Joy, Dec. '79 female

Holly, Dec. '83 female

Nipper, Nov. '87

Beautiful

B.B., Dec. '75 male

Goldie, Dec. '81

Crooked Fin

Puck, Dec. '76 female

Regular beach visitors, 1987: Holey Fin, Holly, Nicky and Nipper (new calf); B.B., Puck; Snubnose: male; Sickle Fin: male.
Irregulars: Crooked Fin; Joy. Another 130 identified dolphins visit the beach at least once a year, out of a total population of some 200 dolphins in the area. Beautiful and Goldie were no longer seen during 1987.

1984 — In May 1984, Wilf Mason wrote telling us Holey Fin's third baby, Holly (Dec 1983) was especially friendly, having allowed people contact at six months — the youngest ever. He advised us that the State Government had financed construction of a Dolphin Welfare Information Centre near the jetty. The building would accommodate a warden's office, audio-visual theatrette, library, studyroom for visiting scientists, and fish-freezing facilities to provide a well-selected store of dolphin food for sale to the public.

During the school holidays people crowded the beach, all wanting to interact with two dolphins. The Foundation has established a no-boating area, restricted netting and placed a speed-limit on boats.

Chi-uh made another visit. After six years of observation she could draw some interesting conclusions about the varying patterns of 'child' rearing by the beach-visiting mothers. Beautiful's offspring, BB and Goldie, born six years apart, were much more independent of their mother and had less to do with the other beach visitors. Holey Fin had a fairly close relationship with her three offspring, Nicky, Joy and Holly, and of all the mothers, she took the most disciplinary action with them. Crooked Fin continued to associate closely with her only calf Puck, aged five. Nicky was the most constant and intensive visitor whereas Joy (4½) who had never accepted fish or hand strokes from humans, now seldom came in. Yet, during Joy's first two years, beach visits had been regular and the calf had spent much time amongst a forest of human legs.

In April a sensitive Australian woman, Alison Celty, gave us an account of her visit which involved an unusual acoustic exchange with Holey Fin. 'I had been watching Jim blowing different noises into the water, to stimulate the dolphin's interest. When she came over to me I knelt and did the same. She made underwater sounds. I listened and tried to copy the pitch and frequency by blowing underwater. Between "conversations" she would lift her head out, nuzzle my face and taste my hair. When she finally parted with the low frequency sound she subsequently used for farewells, I was quite euphoric.'

Later, when Alison's two-year-old son began to drift out to sea on his inflatable raft, 'Holey Fin was there in a flash, behind the raft and followed it right inshore'.

On another occasion Nicky came early in the morning to meet Alison and her friend, Jessica. For an hour and a half the dolphin draped herself around their prone bodies rolling on her side and raising her pectoral fin, nuzzling and tasting them in an intensive and sensual exploration.

In June, Jerry Doran from Hawaii visited. Like so many of our informants he subsequently came to our home in New Zealand and told us how he had spent much time swimming with Nicky and Puck. For them, eye contact seemed most important. He would make underwater sounds to them and saw one make the bubble-gulp gesture in response. He thought it was laughter. Jerry was most impressed with the young couple acting as wardens, giving talks to the tourists and promoting the dolphins' welfare.

1985 — In April 1985, Chi-uh wrote to tell us of her seventh visit. The Information Centre had been completed and they were seeking good educational material for it from all over the world.*

Chi-uh noted that people were swimming with the dolphins much more now and occasionally were even allowed to rest a hand on their flanks as they circled. Two had taken to 'kissing'** people; children discovered that, if they put their hands behind their backs and stooped low over the water, Holey Fin or Snubnose would surface and touch them with their beaks.

At Easter, 250 people crowded the beach. Four dolphins came in and mingled intermittently with throngs of windsurfers, swimmers, waders and fishermen. One man, finding his dog reluctant to swim with the dolphins, threw it on top of one. Photographers crowded each other and cursed. At times the dolphins withdrew offshore for 15 minutes or so, before returning. There was none of the quiet bonding that had been going on prior to Easter, but the dolphins still *chose* to interact. When things settled down again after Easter, the dolphins did *not* resume as before. Beforehand, there had been four or five visiting the beach nearly every day but for two weeks after the holiday, there were only four days of long, intimate encounters, and on five consecutive days the dolphins stayed only ten minutes before leaving.

During those hectic Easter days, Nicky bit several people hard enough to draw blood and thrashed her tail against people's legs to cause bruising. Such incidents occurred when crowds completely surrounded her and the message seemed to be 'let me out'. But Nicky did not depart. She stayed among the

*Donations to Dolphin Welfare Foundation, Denham, Shark Bay, Western Australia, 6537. Membership $15.

**Now discouraged to avoid infection.

swarms of people. Then, a few minutes later, she would be biting and thrashing again. Chi-uh was afraid somebody might strike back in anger.

For three years wardens had been maintained on the beach, paid from slender funds collected by donation. But the need for a fully paid warden skilled in crowd control was urgent if the situation was to continue to evolve.

Chi-uh noted that there were three categories of people who interacted with the dolphins. First, the fishing people, who often saved a fine fish from their catch for the dolphins. Such people had a basic, practical relationship and were responsible for initiating the Monkey Mia situation. Many families had been holidaying there for years and had warm feelings for the dolphins even though they could only name a few and rarely bothered to pet them, other than a fond rub on the head as a dolphin accepted a fish.

Then there were the 'pilgrims', people who, since 1980, had been coming from half a world away, a few at a time, and would spend hours with the dolphins, day after day. Few, if any, fish were given. Eye contact and touch were important but the relationship was one of tranquillity and profound communion.

An example is Raphy Vigod of Canada (March 1985): 'I'm given to keeping my hands behind my back. I give no fish. This morning as I was doing Tai Chi movements, Nicky watched me closely for a long time. At another nice moment one-and-a-half-year-old Holly nuzzled my hand. For the most part I am content to be near them and just let things happen.'

The third category was the tourists, often people passing through on the sealed highway from Perth to Darwin, who called in to pet a wild dolphin or photograph their children feeding them. Clearly the dolphins enjoyed this situation too, even though at peak holiday periods it tended to get out of hand.

With up to 300 people seeking to touch three or four dolphins, they still stayed, sometimes for hours — especially Nicky and Holey Fin. They were chased up and down the beach; people stepped in front of each other to take pictures. Often they tried to control the dolphins as they would their dogs — 'make it lift its head higher', 'stupid animal, it turned away!' (refusing a frozen fish), 'get it to come in shallower', 'Oh, you let it go away'. A man tries to pour his beer down a blowhole; another offers a lighted cigarette; kids throw sand in dolphin's eyes; a man scrapes his finger nail over dolphin skin and inspects his nail to see what it is like.

In such situations the dolphins were moving constantly from person to person, up and down the beach, accepting or rejecting fish. Crooked Fin, BB and the other dolphins would usually stay only a few minutes before withdrawing to deeper water. With so many people seeking attention the mood was one of excitement and frenetic activity, rather than the quiet, intimate atmosphere when there were few people around.

Chi-uh wrote: 'Only after some quiet times with a dolphin do we enjoy letting it be in charge of the encounter, and see what it may decide to do next with us. The notion of allowing a wild animal to be *in charge* of the way we relate to it, seems to be a "blasphemy". Humans are *always* in charge. But perhaps we are at a stage of human development where we could relax and allow a gentler relationship with our fellow creatures on this planet. Those who have tried this approach with dolphins find it most rewarding. I believe this may be what the dolphin/human relationship is all about at Monkey Mia.'

Later in the year Wilf Mason got a surprise. Twice in the same week the dolphins herded schools of large snapper into the beach where visitors could catch them by hand.

And international singing star John Denver flew in on a chartered plane to see the dolphins, which he had heard of from afar.

1986 — By 1986, the once remote caravan park at Monkey Mia was being described as one of Australia's leading tourist attractions and it was included on the itinerary of tour buses. Several films with international circulation had depicted the scene there and the dusty dirt road over the desert from Denham now carried an average of 176 vehicles a day. Within two years the way to the dolphins would be fully sealed.

Hazel Mason told a journalist: 'If we crowded around the Queen when she approached and all wanted to maul her and touch her, she would feel intolerable pressure. If people stand in line here, as they do for the Queen, and wait for the dolphin to come to them, the weight on the animal is removed.'

And so, at peak periods, up to 300 tourists stand in line, many clutching gifts of fish, waiting for an audience with a dolphin. On one occasion Hazel saw Nicky swim slowly along the line at high tide, eyeing the people, then stand on her tail at the end of the file.

In early February our Puerto Rican friend, Alfredo Cabrera, wrote of the hectic situation: 'I worry about the level of activity around the dolphins. There are boats plying in and out all day long and the pier bristles with fishing lines. Holey Fin has lost half of her right tail fluke to a boat propeller. . . I had a most rewarding interaction with two-year-old Holly. In my second encounter Holly accepted a piece of seaweed from my mouth. I hope that when the holiday crowds ease I may have some quality time with her. The behaviour of the people is as interesting as the dolphins. There's a couple here who came to give birth in the water with the dolphins . . .'

Fred and Holly — Dr Fred Donaldson is a play therapist who has spent 15 years working with problem children. From this work he learnt that play is vastly different from contest behaviour and that the essence of play/touch is communication of trust. He then went on to test this theory of play as a form of interspecies communication, with wolves, coyotes, foxes, elk and bears. For Fred the key to such exchanges is in the initial message: 'The intent to play is communicated initially with the eyes,' he wrote. 'The *play look* is not a stare, not an intense probing, but rather a mere glimpse in passing. This "look" happens so fleetingly players receive it simultaneously with its sending. A sense of trust is transmitted and received so quickly, yet this sharing of glances has allowed me to spend many hours in close company with elk, bison, deer and moose, in the high country of Wyoming and Montana.'*

After a remarkable series of adventures with these North American creatures, Fred set out in March 1986 for Monkey Mia. He hoped to see how child and wolf play compared with dolphins. Then, as a result of reading *Dolphin Dolphin* and corresponding with us, Fred sent me his Monkey Mia journal.

His fortnight with the dolphins gradually built up to the hectic Easter period but initially Fred enjoyed days when there were few on the beach. Every day of his visit the same quartet of dolphins came in, with up to nine on some days. Space precludes us from presenting his entire journal here but

* 'Play & Contest in Human-Animal Relationships' by Fred Donaldson Ph.D. Lomi *School Bulletin*, Fall 1982.

in summary form enough emerges to show that a very special bond developed between the 43-year-old man from Big Sky, Montana, and the two-year-old dolphin from Shark Bay. At no time did Fred offer any fish and there were occasions when Holly played with him while fish was on offer nearby.

The Journal

14 March: Arriving at 5pm, Fred hastens eagerly to the beach. The first dolphin to approach him is small with a perfect fin — Holly. She comes within a metre and circles away. Holly then returns directly and nuzzles his fingers. She moves sideways allowing his hand to slide along her flank, brushing his leg with her tail.

15 March: At 8.30am Fred has brief contact with Puck and Holey Fin but Holly behaves differently. Making sounds as she approaches Holly nibbles his fingers. She digs with her beak beneath his foot and nudges it. Then she circles him before heading out to deep water where she flips out several times. . . (an invitation to swim: see 24 March). While Nicky, Puck and Holey Fin swim by permitting touch, Holly comes alone and stays close for long periods.

16 March: From 7.15am until 4pm, the four dolphins stay on the beach. Only Fred and another man are present. While Puck swims between Fred's legs, Holly is still the closest and touches most often.

17 March: Holey Fin and Puck are in at 6.45am. They follow Fred as he walks up and down the beach in the shallows. At 7.15, Puck and Nicky arrive. This day all four stay close to Fred, especially Nicky and Holly. Nicky likes to swim slowly around him, keeping eye contact. Holly enjoys touch. Today is very hot — 120°F. When five sharks arrive beneath the pier, four dolphins herd them out to deep water.

18 March: Puck comes in (7am), then the other three. This morning there are around 30 people on the beach taking photos and feeding — a frantic atmosphere develops. Fred stands back. In the afternoon he kneels chest deep; Holly approaches and he puts his hand under her beak. She stays for a minute making sounds as he talks to her. Later, as he walks along the beach, Holly approaches. Again he kneels as she curves slowly around him. He strokes her chin for a minute.

19 March: The usual four are in by 10am. Nicky comes within centimetres of his face at water level and he rubs her chin. At noon the heat drives both people and dolphins away.

20 March: The quartet arrive at 7.30am and leave at 4pm. Holly comes close and Fred strokes her chin, jaw and teeth as she opens her mouth. Later they play the 'run along the beach' game. Fred is knee deep, his hand on Holly's side. Occasionally she cuts in front, causing him to stumble or jump over her. She then waits and continues. They make several such runs. Near the pier Holly dives and fetches something. She flips a small shark's head to Fred. He tosses it back. For around ten minutes they play catch with it. Then she swims sinuously around him, his head at surface level. In the afternoon Nicky frisks with Fred, circling and leaping out. When he claps she rolls on her side exposing her belly.

21 March: From 7am to 6.30pm the dolphins come intermittently. In the afternoon only Holly comes to Fred and stays close as they walk the beach.

22 March: This day Fred stays in the water from 7.30 to 5.0pm. A regular visitor called Nikki, shows Fred a game she has with Puck. Then Fred tries it with Holly. He holds a piece of seaweed in his teeth and she takes it, touching his lips with her beak in a 'kiss'. They play this game for nearly an hour. Then the 'up and down the beach' game, with Puck and Holey Fin joining in, but only Holly will swim close, Fred's hand on her side. Again she goes out, finds the shark's head and tosses it to him.

23 March: From 7.30, nine dolphins come in, the usual quartet plus Snubnose, BB, Sickle Fin and two others. Lots of people. Holly plays the seaweed game and then, up and down the beach. The male, Snubnose, joins in and this time Fred rests a hand on his side, too. As the number of people increases, Fred withdraws. The dolphins stay until 5.0pm.

24 March: At 7am the four regulars are in. As Holly plays the beach running game with Fred, she seems to want him to go into deep water — like the day she nudged his foot. But Fred, a 6 foot 4 inch mountain man, is afraid of the ocean and sharks and sea snakes. In the afternoon, Holly brings him and Mike Kerswell, the warden, gifts of fish. At the day's end a dolphin gives a fish to an unsuccessful fisherman on the end of the pier.

25 March: At 6.30am the regular quartet arrive but show little interest in people — just passing through. By midday the beach is deserted in the intense heat.

26 March: From 7am to 5pm, Holly and Holey Fin come in and out with brief contact periods. As the intensity of people increases, Fred's contacts decrease.

27 March: At 7am, the four regulars are in. On his last full day Fred spends the morning playing the seaweed game with Holly. 'Today she let me hold her in a big hug without moving away. Puck gave me a fish.'

28 March: At 7am, Puck and Holly come in. Fred has time for a farewell touch. As he says goodbye to Holly she swims in a tight circle around him, looking up. Then she darts out to sea and leaps and twists six times from the water. Mike, the warden, says, 'You know what she's doing, don't you? She's saying goodbye to you.'

1987 — Year by year, the high summer period, when beach visits became irregular, has diminished. Theories to explain the dolphins' withdrawal are hot weather, low water, food supply, breeding etc. Once, the interval extended from mid-November until February. In 1986/87 there was no interruption at all, confirming Jim Hudnall's opinion that beach visiting was really dependent on the presence of people. The hot, windy season was not popular with fishermen but, as dolphin publicity increased, their presence was maintained by tourists.

When our Canadian friend, Marna McDonald arrived there just before Christmas 1986, activity was in full swing. Marna enjoyed the seaweed game with Nick, Puck and Holly. A dangled morsel would be taken away, paraded about and returned. Holly would find plastic bags or a hankie, frolic with them and bring them in, to be discarded.

During her visit, Marna witnessed the first regular acceptance of fish by the male, Sickle Fin. Prior to this he had been a beach visitor but very rarely accepted fish. Now, while waiting around for a handout, even nudging the backs of hands, Sickle Fin made it clear he *did not want to be touched*. Initially he withdrew, but eventually he began to slap his tail and pectorals on the water to emphasise his displeasure.

During 1987, Sickle Fin became a regular on the beach, according to warden Sharon Gosper, who describes this male as a rough, boisterous dolphin who 'makes our job a lot easier as he keeps the crowd pretty much under control'.

In November 1987, we received a visit from Debra Glasgow, a young New Zealand woman who had spent lengthy periods in the last three years working as a research assistant at Monkey Mia, studying and photographing the dolphins for a future book. Debra told us of the most recent discoveries made by the American research team. Some 200 individual dolphins have been identified in the area and many fascinating behaviour patterns are emerging. Puzzling is the habit of certain individuals (non-beach visitors) who carry small sponges around, deftly balanced on their beaks, occasionally in their mouths. Equally fascinating is the fact that it appears a trio of males will sequester a female and keep her in their company for several days, curbing her flight with a curious 'bop,bop,bopping' sound. Perhaps this is related to the estrus cycle. Once, another female was successful in 'rescuing' a captive.

Debra said that swimming and diving with the beach visitors was now commonplace. Quite clearly the dolphins enjoyed it and often prodded people's feet to entice them out. Males would become sexually aggressive toward swimmers. On one occasion she and another woman were induced to accompany a group of dolphins so far offshore there was concern for their safety. Another time, when she felt threatened by an aggressive fish, Puck left a group of people with fish gifts and came to her aid, driving it away.

The dolphins showed especial interest in pregnant women, scanning their abdomen as they sat in the water. They were equally fascinated with small babies. Speculation was rife that Nicky, now in her eleventh year, might be pregnant . . .

Advice to Visitors — An Australian researcher, Geoff Hume-Cook, has been making a video film-study of human/dolphin relationships on the beach at Monkey Mia, trying to establish why humans are so fascinated by dolphins and what they expect from such contact. Geoff offers visitors excellent advice:

'The people who seem to relate best are those who let the dolphins make the first move or those who have taken the time to build up a trusting relationship with them. Gentleness and sensitivity are very important. I've seen visitors go down to the beach for the first time and get quite disproportionate amounts of attention. Others can wade into the water brandishing fish, but when they go to touch an animal they are firmly rebuffed.

'A lot of people seem to see them not as living creatures, but as a tourist experience. But what they're really dealing with is a highly intelligent wild animal we are still trying to understand.

'Many people have a centuries-old image of dolphins as friendly creatures with a fascinating affinity for man. At Monkey Mia you can hear them say how sweet or clever the dolphins are when they seem to smile, frolic or turn on an exhibition of swimming and leaping. What they're actually seeing more likely is exasperation, sexual play or rowdy social behaviour as the animals get on with their own lives.'

Geoff has found that quiet people who get along best with dolphins were often unaware of the subtleties of their approach. His film shows that communication is more often conveyed through physical and non-verbal expressions; rarely through sound or words.

'You can be too eager to be friends. You have to give them time to size you up. You should never reach across a dolphin's head and you must never try to touch its nose or blowhole. They also dislike being touched on the tail or dorsal fin, and they are wary of anyone who makes sudden or inappropriate overtures. Dolphins can show their annoyance by a slight shudder the length of their bodies or, if they have had a really bad day, they can bite.

'If visitors could understand how privileged they are just to be able to watch these animals at close quarters they would spare them a lot of pressure.'

New Birth — During Debra's visit, a letter arrived from Sharon Gosper: 'We had an exciting past few days at Monkey Mia. Nicky gave birth between 18-20 November. She brought her calf in on Saturday and again yesterday (24/25 Nov). It is absolutely beautiful, about 70cm long. Nicky came close and took several fish from my hand, careful that her baby was sheltered on the other side. Twice it came under her body and dashed in amongst the people's legs. Nicky quickly collected it and swam out to sea. It is incredible that she is so trusting.'

Twelve years earlier, in December 1975, Nicky herself had been brought to the beach by Holey Fin, now a grandmother. The story of Monkey Mia is a continuing one and much more will be learnt as research is published. But with Nicky's calf, an historic cycle has been completed; a free dolphin, habituated to humans from birth, has brought her new offspring to the interface.

The Maravilla Dolphins: Bahamas

The history of encounters with solitary dolphins and small groups, such as at Monkey Mia, offers valuable insights into ocean mind. Clearly, there is a potential for interaction with our species. But how far might this extend if we could meet dolphins with the same degree of intimacy *within their normal social group*? Our chapter on chance diver/dolphin encounters shows there is an intensity of interest between species but the situation is usually too ephemeral for more complex relationships to evolve. The stories of Donald, Jean-Louis, Sandy and the others show that deep bonds do not develop by magic. As with any friendship regular contacts lead to familiarity and trust. For this to occur between oceanic and terrestrial creatures, certain physical handicaps have to be overcome. Daily human access to free-ranging dolphins requires that the dolphins are handy to a regular food supply and yet sufficiently remote from human population pressure to avoid pollution, overfishing, harassment and other negative human input.

Few areas in the world offer such conditions, and knowledge of them entails new problems for the dolphins. Until our species accepts the etiquette any healthy relationship demands, the full potential for interspecies friendship will not be realised. But already some important growth *has* occurred, and if there is to be any advance in understanding, it is probably better to share this knowledge than to keep it under wraps.

Wreck sites and places where people engage in undersea tasks often feature on our files as places where dolphins have turned up regularly to observe human activity. Such spots are often isolated and so demanding of the divers that they cannot down tools and harass the dolphins with the stock 'grab a ride' response our species manifests when dolphins are sighted. At best, when we see dolphins we stop whatever we are doing to stare in wonder, little thinking that to the dolphins we soon become dull. But treasure divers, it seems, are a special case . . .

The friendliest group of dolphins in the world live near the wreck of a famous treasure ship. To me that seems no coincidence.

Three centuries ago, human tragedy entered the ocean territory of a tribe of spotted dolphins in the northern Bahamas. In the darkness of 4 January 1656, an armada of 14 ships bearing a great treasure to the king of Spain, was clearing the dreaded Bahama Channel, heading east across the Atlantic for Cadiz. Around midnight disaster struck. *Nuestra Senora de la Maravilla*, the lead ship, suddenly found herself in shallow water over a sand-bank. A warning cannon was fired. In the confusion there was a collision and the *Maravilla* sank rapidly in 16 metres. A violent storm ensued and of the 650 people aboard, only 45 were rescued. Spanish sailors recovered a little of her $850 million dollar hoard before she vanished beneath the shifting sands.

In chapter 12 of *Dolphin Dolphin*, I describe at length how, in 1979, we came to meet film-makers Michael and Morgan Wiese, and learnt from them of the friendly *Maravilla* dolphin tribe. As a result I was to travel twice, half across the world to meet these dolphins and by a stroke of luck, I even encountered the treasure hunters with whom they first established rapport.

Back in 1962, Bob Marx and Teddy Tucker began searching for one of the world's richest treasures. For a decade they towed wreck detectors and their own bodies, up and down over the vast sandy plains that long ago may have constituted the most northern island in the Bahamas. On ancient charts the area once featured as Tumbado Island.

Bob Marx would be one of the world's foremost treasure hunters. Author of 31 books and knighted by the Spanish Government for his work in marine archaeology, Marx first became fascinated with the story of *Maravilla* in 1960, while researching in the archives at Seville. There he came upon a book published by Dr Don Diego, one of the wreck survivors, with charts revealing her precise location.

After searching an area of 50 square miles on the sandy plateau and discovering 20 other wrecks in the process, Marx located the *Maravilla* by a stroke of luck. In August 1972, the anchor of his salvage vessel *Grifon* caught on the galleon's ballast stones. With the ship's powerful water blaster — a metal tube that enshrouds the propeller, deflecting the wash down into the sand — a crater was excavated, eight metres to a hard limestone bottom. And then the fun began. By sundown the divers had recovered a beautiful gold dish in the form of a scallop shell, silver cutlery and two sets of brass navigational dividers. The next day they found four iron anchors and two bronze cannon. By the fifth day they had 2000 silver coins and five silver bars weighing 70 pounds a piece. Two months later they had two million dollars worth of treasure.

Over the ensuing years Bob Marx and his men have worked the wrecksite for long, non-stop periods, observed by a tribe of about 200 spotted dolphins. For the divers, these creatures became a welcome sight. Working ten hours a day underwater, along with an iron man like Marx, guarded by men with guns, the task of treasure recovery was no picnic. But Marx has a soft spot for dolphins. Mention them and he gets tears in his eyes. Dolphins probably saved his life.

Towing behind the ship on what divers ghoulishly call the 'sharkline', Bob saw two dolphins heading at him on a collision course. Something made him glance back. A large hammerhead shark was closing on his fins. That instant, the dolphins rammed the shark and next moment several others were fending it off and driving it away. No wonder Bob Marx has a huge regard for dolphins . . .

Whenever the dolphin tribe came over the wrecksite, it became playtime. The dolphins were the divers' only rest and recreation. Each was an expert in the water and the dolphins must have found their first group of humans superb to gambol with. When I met Mike Daniel, one of the chief divers, he showed me underwater photos dating back to 1973, and Marx told me of movie footage he had shot in 1974.

So, when film-makers Michael Wiese and Hardy Jones came down to Florida in 1978, after searching in the Pacific and Atlantic for friendly dolphins, Bob Marx directed them to the area on Little Bahama Bank, where his treasure ship lay.

In June, when the weather is most promising, they set sail on the 22-metre schooner, *William H. Aubury*, for the Maravilla site. They encountered a group of about 60 Atlantic spotted dolphins, *Stenella plagiodon*. They played with them for three days for periods of 45 minutes to three hours. Each session ended when they were too tired to swim any longer.

PLATE 21A

During a 1984 expedition to the Bahamas, spotted dolphins arrive escorting a swimmer, virtually 'carrying' him in their wake. *Rosemole* is on the left.

PLATE 21B

Michele Daniel, just five years old, aroused intense interest when the dolphins encountered their youngest human, inducing to dive for the first time.

PLATE 22A

As skipper Bill Griffin dives to pick up his camera lens a dolphin pair adopt a most unusual pose suggesting a lenticul
pattern.

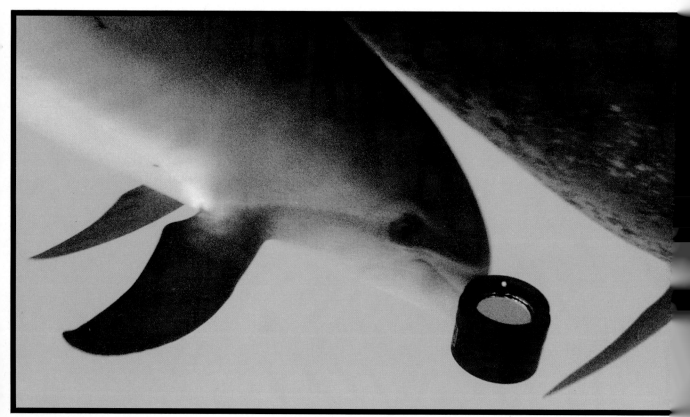

PLATE 22B

As the lens tumbles down, the dolphins swoop in, nuzzle it, let it roll along their backs, and frolic with each other.

Credit: Mike Daniel

snorkel, or any human possession, deliberately dropped produces similar responses: curiosity and patterned gamesplay.

Credit: Mike Daniel

the close presence of humans these spotted dolphins frequently caress and nudge each other. The young of the species
e no spots.

PLATE 24A   *Credit: Jack McKenn*

Frequently, while interacting with people, love-making and foreplay are exhibited, as if a deliberate gesture, an aspe of play.

PLATE 24B   *Credit: Jack McKer*

At the outset, the dolphins rode the bows of the schooner, while 11 men and two women crowded around the bow-sprit. Then Hardy Jones and cameraman Jack McKenney, clinging to the chain plates, dangled their bodies in the bow wave among the dolphins to establish mutual trust.

Although the film shows a different sequence, it was a flute that initiated the interlock. When the ship anchored, two hydrophones were suspended beneath her keel over the shallow ocean floor. Through them, Mary Earle played 'Apalachian Spring' on the flute. To their delight, the dolphins returned to the ship and people began to enter the gin-clear water. One of the most successful at relating to the dolphins was Dr John Siebel, an experienced snorkel-diver who could hold his breath a long time. He imitated the motions of the dolphins and would often hum to them.

The interlock developed into an elaborate and graceful interspecies ballet when sound engineer Steve Gagné drifted down cradling in his arms a pneumatic underwater piano. Using the exhaust of his scuba, Steve had devised a keyboard instrument which produces high-pitched crooning notes.

Morgan Smith was dolphin-kicking when a dolphin swam above her, paralleling her movements before gently touching her on the back. Humans and dolphins cavorted in perfect conditions; water that was extremely clear with more than 30 metres visibility, 83 degrees warm and only seven to ten metres deep — and a limitless white sand bottom. Only the two cameramen and Steve wore scuba but the dolphins showed no fear of them.

On the second day contact began at five in the afternoon. Michael Wiese told me: 'Three of us were on the bow. Beautiful weather and light. We said, "It's so perfect — now if only we had dolphins." Seconds later the ocean was filled with leaping dolphins, heading to our ship to ride the bow wave.

'Down below certain dolphins seemed to pick me out of the bunch of humans and I felt that if we had been able to remain on location longer, individual relationships might have developed.

'On the third and final day, the dolphins were summonsed by Steve's underwater piano. At first, about six to eight dolphins came in, almost like a scouting party and then others arrived. About 20 minutes later they left, apparently to feed. We could see birds feeding a kilometre or so away in the direction they headed.

'That evening the dolphins hung around the ship and watched us. They played games with one of our rowboats, twisting and pulling on the tow rope, tangling the boats up.

'The most powerful communication to us from the dolphins was this: we had spotted some very large and nasty-looking barracuda one day while swimming with the dolphins. They obviously knew of our fear because occasionally they would chase the barracuda away. The next day, in one of the motorboats towed behind the schooner, we found a stiff, dried out and very dead barracuda with curved tooth marks in its back. The dolphins seemed to have tossed it up like a gift during the night as if saying "hurumphff! Barracuda — nothing to be afraid of. Here".'

Following the success of this filming expedition, Hardy Jones made two return trips before his team re-established contact in June 1980, after an eight-day search, and filmed a sequel to *Dolphin*.

This time they chartered the sailing vessel, *Sir Cloudesley Shovell*. To help pay expenses, Hardy Jones included dolphin-loving people who would pay a share of the voyage. Among them was a young San Franciscan woman, Edith Howland, who later became our close friend.

On June 10, the ship was anchored on the Bank in ten metres of water and lunch was in progress when dolphins arrived. A mad scramble and most of those on board were in the water. Edith found herself almost numb with excitement.

'As they came in I could hear sounds getting louder and soon came to realise I had only to look around once I'd heard them, and there they would be. Approaching head-on, they would scan me with sonar, heads swaying to and fro, emitting rapid click chains at me. They were very close and appeared suspended in the intense blue, like a vast pool with endless vistas. The clarity and warmth (83°F); feelings of love, respect and exhilaration — so many thoughts racing through my head. The glint of light on their steely grey skins; their size; their eyes seeking contact; the way a trio swam in perfect line abreast. Gradually more came in until there were about a dozen. Such economy of motion. They were especially close in a game where I swam fast behind them and then stopped. They would hover just in front of me, glancing back. I found they came closest when I kept my hands at my side. At first I held them out, thinking they might want to peruse these odd extremities. They had astonishing control, maintaining a fraction of an inch from my hand by a muscular quiver that was plainly visible. This seemed to come from an overall body awareness rather than eyesight. Eventually I felt a little chilled and very salty, but couldn't bear for this to end. The encounter lasted one and a half hours.'

During the eight-day expedition there were three more dolphin visits: one at dawn and two at dusk, each lasting about an hour. Hardy Jones noted some changes from the 1978 meeting: no longer did the dolphins seem attracted by music, but came and went at will. 'They habituate to novelty very rapidly,' he commented.

With dolphin behaviourist, Karen Pryor, on board, they were able to identify 16 individuals of the 28 they sighted in all. At no stage did they see the whole group of 60 but began to distinguish broad relationship patterns. There was a juvenile sub-group, ages two to five, rambunctious and playful, that swam together, tumbling, leaping and jaw snapping at each other. Another sub-group was three mothers with babies, and a third, adult males.

When the dolphins investigated a camera lying on the bottom, Hardy decided to drop one of his swim fins. Five dolphins responded with buzz trains of sonar as they hurtled towards the spiralling object. One went right in and touched it with her beak. This particular dolphin emerged as a dominant personality. Nearly always accompanied by four juveniles, she would approach divers closely, emitting a recognisable signature whistle, 'di-di', with a stream of small bubbles from her blowhole. Didi, as she became known, had another distinction: a remora or suckerfish was invariably sliding around on her body. One of her group, believed to be her offspring, had the tip of its dorsal missing — Chopper.

In June 1981, Hardy returned, this time with a film crew, special sound apparatus, and a bigger vessel, the 30-metre *Tropic Bird*. When they dropped anchor in the dolphin zone, Hardy and three other divers leapt over. Five dolphins approached. The closest reminded him of Didi, but appeared larger and more heavily spotted. As a test Hardy dropped his swim fin. 'The dolphins dived and wheeled towards me in mid-water as my fin floated down. One emerged from the pack, turned upside down and approached within inches of the fin. As it turned I saw the remora. It *was* Didi, still accompanied by her hitchhiker.'*

For 35 days the film expedition remained on site. As in the previous year, the adult male sub-group, 'the heavies' as they were called, remained initially aloof. Then little by little that slow, deliberate squadron began to mingle freely with the divers, once for 45 minutes. They found the dolphin visits most frequent at dawn and dusk, and believed they rested between ten and four, feeding at night out in the adjacent Gulf Stream.

The most interesting interactions involved the new experiments sound-engineer, Steve Gagné, had devised. First, they played a previously recorded tape of dolphin sounds. The dolphins approached with excitement, milling at the surface, orienting at the underwater speaker. Then the adult males arrived and submerged. At the bottom they clustered motionless around the device while the juveniles began to play sexual/aggressive games — tail-slapping, biting, jaw clacking, back flops, and belly up swimming.

Over the next three weeks the expedition concentrated on filming in conjunction with Steve's special recording apparatus. Synchronised sound and images would enable them to correlate the dolphins' vocalisations with behaviours.

In the fourth week, Steve introduced his sound generator. Inspired by Didi and her signature whistle, he had constructed a computer-driven synthesiser which could produce ten different signature whistles. Filmed by Howard Hall, with Steve recording, Hardy tried the device as the dolphins circled him. Hardy thought he heard the whistlecall twice and feared that the button had jammed. Then he realised — the dolphin had mimicked the sound he made. He ran through the entire repertoire of ten calls. Some were copied, others ignored. Call number seven sent all dolphins scattering. He would not use that again. But call three appeared to attract them. As Hardy continued to generate the whistlecalls, the dolphins began to add something new in response. Not just mimicry but an embellishment. He felt frustrated, stuck with his one note while they could improvise like jazzmen.

Then their swimming pattern changed radically. A male called Jagged positioned itself at the surface and began sinking tail first until it was 'standing' on the bottom. It then lay belly down on the sand before slowly and deliberately coming to the upright position, flukes resting on the sand.

Months later, reviewing this extraordinary film sequence, the three divers came to appreciate what had taken place. The dolphin had copied their vertical descent with the equipment — movie cameras, tape recorder and sound generator. Then it mimicked their actions on the bottom while they were operating it. Hardy felt that, in a primitive way, they were really conversing. The interlude concluded as the 'heavies' started coming in.

Biologist Julia Whitty, Hardy's wife, felt this was what behaviourists call 'experimental mimicry', duplicating an action to see how it felt and perhaps what it meant; something that children often do, along with apes and chimpanzees — 'but this is not the kind of thing you'd ever see a dog do . . .'

A few years later, radio astronomer, Gerrit Veerschuur, was to see this film sequence on television. He was transfixed. Gerrit operated the world's largest radio telescope in Puerto Rico and was a member of an illustrious

*The third time Hardy did this, the following year, a dolphin defaecated on his swim fin.

group of scientists engaged in the quest for alien intelligence: the SETI project.

For many years these people have studied the farthest reaches of the Universe in hope of locating some signal from an intelligent being. According to the mathematics of probability, the likelihood of our species being the only sapient life form in Universe is absurd.

At their annual symposium, held at Greenbank, West Virginia, in May 1985, Gerrit told his colleagues:

'The film, The Kingdom of the Dolphins (which I saw on PBS), revealed an astonishing example of interspecies mimicry which could provide a lesson for us all. The humans involved had built a synthesiser which generated specific whistles similar to dolphin signature whistles. On hearing this, one dolphin, who seemed to be the ambassador for the group, responded by mimicking the sound. The dolphin then added a footnote, a further comment, if you will, hoping (or perhaps expecting) that the humans would, in turn, mimic her and communication would begin. However, the humans were not prepared for this and would not respond, due to severe technological and physiological limitations. Perhaps they hadn't expected the dolphin to be 'so intelligent' as to respond immediately and so profoundly. The dolphin did not give up, however. As an additional signal that she was aware and conscious of the human's effort to communicate, the dolphin mimicked the prone position of the cameraman lying in the sand and then oriented herself into a vertical position, tail in the sand, in front of two other humans who were standing on the sea floor. It is difficult to imagine any other way for a dolphin to mimic a human except through sound and bodily position or movement. So was contact made? Sure. But who was the intelligent one, or the more conscious? Can we even begin to prepare ourselves for such extraordinary attempts to communicate with us?

'We will first have to mimic any signal received from E.T. This simple form of response will inform them that we are here, and are willing and able to duplicate their signal. We will surely add something to spice up our response, just as the dolphins did in that film. We will hope that E.T. will deal with our responses better than we dealt with the dolphin's efforts.

'What we add to the returned message is a major question. But are we psychologically prepared? What will be involved? We will, no doubt, some day assign a considerable amount of human energy to this question. A reply will require heavy-duty transmitters, large radio telescopes and high-level committees to do the job, which again touches upon the issue of *who* decides what to say.

'We will probably be confronted with the unexpected. Mimicry may be part of contact and the dolphin connection is worth further exploring for the insights we may gain from experimenting with this form of interspecies communication. To me the dolphins are a form of ETI and provide an incredible opportunity to test our computer technology and our decoding algorithms, if we wish to do so. The cetacean species has demonstrated a willingness to communicate and highlighted an amazing inability on our part to deal with this.

'Twenty-four years ago, at the first SETI meeting in Greenbank, they created The Order of the Dolphin. I would like to re-establish contact with the dolphin awareness. I have written a proposal on how to make contact with the dolphins. You need the best learning program people, and the best decoding experts, build a learning machine in a small box which we submerge to literally swim with the dolphins and let the dolphins teach the machine, essentially a new-born dolphin, and therefore us, how they communicate.'

Hardy Jones and his wife continued to make film expeditions to the Maravilla dolphins. In subsequent years they were able to identify some 70 individuals and they determined the extent of the tribes' home-range, an area of some 40 square miles on the Banks, beyond which they would not venture.

Hardy's films document some fascinating behaviours, especially with Didi, who was to continue dominating the scene until 1984, when she became strangely reticent.

On one occasion, she was filmed picking up a sea cucumber, balancing it on her beak and passing it among the other dolphins. But when Didi took Hardy's tee-shirt and passed it to her baby, Jagged came in, took it and vanished into the blue. The shirt was not seen again.

Didi was offered a plastic windup fish which wriggled its tail and clicked. First she approached very close and ensonified it. Next she balanced the toy on her beak and blasted it with sound. 'Then a playful look came in her eye. She swam down, dug up a sea cucumber, brought it over and offered it to us as a gift.'

This sequence and others, which show prey being captured and barracuda repelled, add weight to the theory of Dr Ken Norris, that dolphins stun prey with sound. (A diver who had a steel plate in her wrist often felt a tingling sensation when with the dolphins. Another diver used his electronic watch to create a signal.)

1983: First Journey to the Bahamas — The wildest dreams *can* come true. I had been following developments with the Maravilla dolphins from afar. Few places seemed more remote from New Zealand than the Bahamas. Then, in 1983, I was invited, all fares paid, to the Whales Alive Conference, held under the auspices of the International Whaling Commission, at Boston, Massachusetts. The last thing I expected to experience on that long journey across the Pacific and America was cetaceans. And yet, after three days of talking about the welfare of cetaceans, the whole conference was embarked upon three whale-watch vessels and taken out to Stellwagen Bank, a whale feeding ground off Cape Cod.

To my utter delight two humpback whales approached our vessel and made eye contact with all on board, all the illuminati from the conference: Sir Peter Scott, Walter Cronkite, Dr John Lilly and all.

When first sighted the whales were feeding on tiny sand lance fish 60 metres below. We silently hove to in the vicinity. The springtime Atlantic was green with plankton. Soon the whales started to do shallow, breath-recovery dives. Then they slid gently below the surface and headed towards us. Soon we could see their white flippers, seven metres from tip to tip waving like giant wings in the green twilight. They approached our vessel beam-on, 100 people thronging the rails, cameras whirring and clicking. As they passed beneath the keel there was a stampede for the other side where they surfaced near the hull. One then slid under and hovered beneath the stern. When it surfaced only a metre away, its eye met ours and its great blowhole pulsated with air — a mass of cameras, an array of lenses, a forest of humans. Cries of jubilation! I have never been with more ecstatic people.

During the conference I was approached by Larry Vertefay, who had been corresponding with me for several years about the Maravilla dolphins. Of mixed Red Indian and French ancestry, Larry's bright eyes glowed with warmth as we met. He had been on several of Hardy Jones' expeditions as a fare-paying passenger. Obsessed with dolphins, he had formed a group called 'Friends of the Sea' and was arranging charters to go out there himself and further his understanding. Down in Florida he had the big dive boat *Bottom Time* all prepared. All I had to do was fly down to Fort Lauderdale. Jim Nollman, the interspecies musician, and his wife Katie, were also invited.

On the morning of 17 June 1983, I woke from a dream as *Bottom Time* danced northward along the Little Bahama Bank at full speed (22 knots). In my sleep I had seen three dolphins approach the bow. I staggered on deck. There they were, an adult and two juvenile spotted dolphins, riding our bow wave. The adult left when we anchored, but the zodiac circled and came in with the juveniles on its bow.

The first spotted dolphin I ever met had very few spots; except for the long beak, she might have been a bottlenose. But on her right flank, just below the dorsal, was an elliptical black mark. I called her Rosemole and felt close to her as she glided beside me, her eye shining. An unforgettable moment, that was to echo through my life.

For half an hour, the dolphins kept everybody happy. There were so many people in the water it was rare to develop rapport. As at a cocktail party, people kept cutting each other out, quite unwittingly. Seeing a dolphin almost stationary, somebody would race up, and that would be it.

In the afternoon the dolphins returned. This time some adults accompanied them, sombre, heavily spotted, almost touching as they cruised around us like a squadron of huge cigars. One individual had scallops out of the trailing edge of the right pectoral and it became Scallopfin. With a dozen people in the water, it was cocktail time again. The nicest contacts were when a dolphin swam parallel and made eye contact but such moments were easily interrupted by an enthusiastic gamester. Several people attempted to touch dolphins, in spite of advice to the contrary. After 30 minutes, the dolphins withdrew. Twice more that day, three dolphins returned briefly.

The next day our vessel moved south along the Bank and met a group of 20 spotted dolphins. With them were Scallopfin and Threadfin from the previous day. Initially, a large number of people leapt in, creating an interspecies melee. It was enough for me to actually inhabit that exquisite world familiar from the film *Dolphin*, but I had not bargained on the delicious warmth — better than atolls that I have dived on the Pacific equator, thanks to the Gulf Stream/Sargasso Sea heat.

The brilliant white sand is pure calcium carbonate which keeps precipitating from the sea water as the sun heats the shallows, rebuilding the Banks as constantly as the storms erode. To meet dolphins in such a world is a perfect fantasy. It was like a vast, mid-ocean swimming pool. The intense undersea light reflected from the sand made everything appear surreal. The dolphins gleamed and sparkled, their undersides lit from below. There seemed to be nothing between them or the sand or the divers — a vacuum of inner space. I had to *remember* to breathe. This was *not* a dream but the closest I have experienced in a waking state. Interlock deluxe! As the dolphins withdrew, the last to leave was Scallopfin.

At four o'clock in the afternoon, a few dolphins returned. Now was the time to try a different approach. Everybody on board had experienced the dolphins and felt ready for a change. Pam Herbert and I formed a pair, interacting with each other underwater, dolphin-kicking in unison and spiralling around each other in a double helix as we ascended slowly. A pair of dolphins had detached and were watching us. Then they copied our actions, undulating beneath us, rubbing and gently twisting. They disappeared, but returned instantly with another pair, who watched us all over again. When we eventually got out after an enchanting time, Pam said, 'It was so right,' and later explained that few actions in her life had ever felt so harmonious — 'simple, unerring and resoundingly sweet. The dolphins seemed to transmit messages that went beyond mere acceptance; there was joy, comradery and non-verbal cheering.' From the upper deck those on board watched this interlude and it was video-filmed for study in the saloon.

Toward the end of the day, the dolphins returned. We had a new game in mind. Pam and I linked up. We reached out for Larry. He joined with his son Dane. The video camera captured it perfectly, as time and again the dolphins came to us in a matching quartet. Another trio hovered nearby — occasionally one tried to join in, but appeared to reconsider. The quartet would approach us slowly, head-on, and peel off either side with slow, steady eye contact. In the quartet were Scallopfin and Threadfin. Earlier Larry thought he saw Chopper from previous years.

Reviewing these encounters on video was of great benefit. We saw that when we maintained a close, co-ordinated group the dolphins could interact with four people without having to break up their own set. The duration and complexity of these encounters seemed to be more interesting to them than our earlier behaviour. *We were being trained.*

The next day, as we headed back to port, we had two more encounters with a different set of dolphins: eight adults, including a mother and baby. The old pattern of everybody in the water at once recurred. We tried but it was not possible to co-ordinate with more than four people. After ten minutes the dolphins left.

The weather report indicated a cyclone was forming in the vicinity. In such areas the atmosphere begins to boil. Water spouts snaked on the eastern horizon. Just 20 kilometres away, the skipper explained, were 160-kilometre-an-hour wind storms. *Bottom Time* headed back to West End and then, in three light planes we flew over the Gulf Stream, which took 40 minutes from Grand Bahamas to Fort Lauderdale. By the end of that day, chasing the sun across a continent, I was in Los Angeles and bound for New Zealand.

Flying over the Pacific, there was time to think over the whirlwind of events I had experienced. In my brief interlude with the Maravilla dolphins we had met, in all that watery wilderness, the same pair on three occasions, in varying company and over a range of many kilometres. This was the only place in the world where people could predictably swim with dolphins in the open ocean day after day. With prolonged contact those sensitive to the needs of the dolphins might explore the subtleties of ocean mind under ideal conditions. Perhaps it would be best if there were scientific observers who would take a passive role, concentrate on identification, document any developments and leave others free to interact spontaneously. This would allow some degree of objectivity, assisted by video cameras and sound-recording gear. On the other hand, communication is an art, not a science, and the presence of wild dolphins for observation depends on human creativity.

My parting words to Larry Vertefay had been that to go any further, it would be essential to have his own vessel, specially fitted out for dolphin research, ideally with the stability of a catamaran, so that people could stay out on the Bank and adapt to the dolphins' living patterns. All I had in mind was a small sailing vessel, rather like our own . . .

1984: Second Journey to the Bahamas — A year later Jan and I flew to Florida. At the airport, Larry and his wife Mary were waiting and they whisked us off to a boatyard to see the catamaran they had bought — *Dolphin*, a British built 20-metre fi-glass/balsa-core luxury sailing vessel, nine metres in beam with 30-metre mast, twin diesels and five bathrooms! The Vertefays may be paupers for years to come, but the dolphin obsession had brought them to this: 'Here's the boat. Let's fit her out for dolphin research,' and so, at their invitation and expense, Jan and I were to have five months in Florida and the Bahamas . . .

Larry's aim was quite simple: he wanted to take dolphin enthusiasts to meet the Maravilla dolphins and document the developments on videotape. As well as video-filming from on deck and underwater, each session would be recorded on soundtape through a hydrophone; and pleasant music would be broadcast from the ship through an underwater speaker — not as a dolphin attractant but to enhance the setting. To assist in documentation, people would be urged to remain as near the vessel as possible. Larry sought a wide range of people: young and old, male and female, extrovert and introvert, athletic or ordinary: all that mattered was *enthusiasm*.

Looking back over our diaries after a series of 33 dolphin encounters, certain patterns suggest that in a subtle way we were manipulated by the dolphins towards behaviours which are not the normal human approach to problem solving. While some of our actions had to be premeditated, or nothing would have happened, the more spontaneous we could be, the more unexpected the outcome. As long as the dolphins were in control, with us happily enjoying the gamesplay, we forgot to be manipulative.

The video cameraman was a special case, however. His passivity may have intrigued the dolphins because they often rubbed his shoulders.* Having the opportunity to review each session topside afterwards, we gained a detached view of our actions and saw events in a larger context. Fresh insights could be readily shared with all on board and newcomers each trip could swiftly gain an overview of progress. In this way the dolphins did not have to start from the beginning and teach us an appropriate etiquette.

While we endeavoured to avoid anything structured, one guideline was important: we would try not to have too many people in the water at one time. To randomise things, we eventually drew straws to decide who would enter first when the dolphins arrived. In this way the dolphins got to meet the shyer, less dominant people on board and everybody had the chance for a quiet tête à tête.

The First Voyage — Five weeks of arduous work were needed to refit the big catamaran for her new purpose. On the positive side, this bonded us closely for voyages together. As skipper, Larry obtained the services of Bill Griffin, an experienced diver and dolphin enthusiast. By a stroke of luck he brought as firstmate, his close friend Mike Daniel, one of the original *Maravilla* treasure divers. Following many setbacks and frustrations *Dolphin* eventually set sail for the Bahamas with 11 people aboard.

Day One (18 July 1984): Shortly after dropping anchor near the day's end, 11 dolphins arrived, including *Didi* who now seemed to keep to herself. For the first time in her life, Jan looked down into the warm, clear Atlantic and

---

* A member of Hardy Jones' team found that if he clasped his hands before him he was often touched.

saw me just above the sand, dolphin-swimming, a dorsal fin attached to my weightbelt, surrounded by a seething mass of dolphin bodies, all of us turning and rolling, twisting and tumbling together. Gracefully the dolphins wove a tight circle around me, so close I was barely visible, their bodies almost touching but never colliding. Mike Daniel was watching too. He turned to Jan: 'That was nice, eh?'

Jan joined me below and we dolphin-swam in unison as courting dolphins do. We held hands and stroked each other. I was proud to show them my partner. Two dolphins circled us tranquilly — gentle and near. They were juveniles with no spots, except one who had an elliptical black mark on the right side. This dolphin looked deep into my eyes. It was Rosemole, the first dolphin I had met the year before.

One dolphin turned belly up and rubbed its back along that of its partner undulating below, the mirror opposite of what Jan and I had been doing belly to belly. Then another pair swam past, belly to belly.

Nearby Patrick (16) was swimming fast. He, too, was enveloped in dolphins, but they were moving at high speed. He torpedoed along, arms folded over his chest and then rocketed to the surface where a dolphin leapt right over him. Mike Daniel took off his snorkel and tossed it out to the dolphins. Several followed it to the sand, nosing it all the way. A group were standing on the bottom, balanced on their tails, actually stirring the sand with their flukes as divers do just before they rise for air.

Larry spent an hour swimming alone with his old acquaintance, *Didi*. Still with her pet remora, she kept about 20 feet away from everybody else.

Dusk was descending and only Patrick, Jan and I remained in the sea. The current had strengthened and I clung to the anchor rope half way down. When I called 'di, di, di', a dolphin swept within a metre of me, making a distinctive whistle, and then thrust its head above the surface to the delight of those on the bows.

Later, as we were drying ourselves on deck in the growing dark, Didi returned and made a magnificent leap against the rising moon — a parting gesture it seemed. Larry told us there was something odd about Didi now. He did see her offspring Chopper. Mike told us that many of these dolphins, including Didi, would have been familiar with divers all their lives. He had been meeting them ever since he started working the *Maravilla* wreck in 1972. In those days, divers invariably leapt in with cameras when they saw dolphins, and some would try to grab a ride. They never came close and his early pictures show large groups of adult dolphins circling the divers. Mike agreed that our 'hands off' policy, and 'no cameras at first', but just spontaneous play, seemed to make a huge difference to the quality of the exchanges. With regard to Didi, he speculated that her distinctive sound may have originated from the treasure divers' habit of calling each other in the water 'do-do-do'. Perhaps this was her version of that call.

Day Two (19 July): This was a repeat of the previous day with the same intensive play, but a *different* set of dolphins. Around 11am with the underwater speaker filling the sea with Bach flute sonatas by Jean Pierre Rampal, two adult dolphins arrived. Larry saw them hovering near the bottom beneath the speaker, their tempo changing with the music, remaining below for extended dives. Three juveniles joined them but were soon drawn to playing as more people entered the water. Before long there were nine dolphins, including four adults. I saw Scallopfin from last year. There was a mother with a new baby that had its own tiny remora: Remora Kid.

Patrick raced past, unaware that a dolphin had its beak almost touching his swim fins. Dane spiralled upward with a matching pair. Casimir pushed the bulky video camera in front of him, a dolphin either side brushing his shoulders as they passed. Dean, with a plastic slate, drew a mother and baby and showed it to a mother and youngster. A companion with them took a close interest in the sketch. Dane and Patrick saw an adult snatch a little razor fish from the sand and swallow it. Eight slender tuna cruised past with the dolphins. In front of Jan, a dolphin opened its mouth and wriggled its tongue. Three juveniles swam past Dane, passing a piece of seaweed from one pectoral to the other and then to Dane.

Just on dusk, Didi appeared alone off the starboard bow, as before. Larry called. She responded, putting her head out and looking at him. She raced around with a scrap of seaweed on her tail, a small tuna following her.

Day Three (20 July): Around 10am, two dolphins arrived including Chopper. A constant companion slightly bigger, had two nicks from its dorsal. Bill and Jan joined them in the water. The mood was gentle and intimate. Jan dived and turned on her back, twisting and spiralling. The dolphins closed around her. Mike tossed his amphibious lens to the dolphins. As it fell they rubbed on it and caught it on their backs, all the way to the sand. Bill retrieved it and Jan watched as a dolphin nuzzled the sand, weaving back and forth as if searching for the lens — until it realised Bill had it.

Nine dolphins enfolded Mike, almost touching him. There was a lot of contact among them today; moving slowly, gently mouthing and caressing as they enclosed Jan in their midst and rose vertically with her to breathe, just as on the first day. On the bottom they were all together, winding around each other as Jan approached. She wondered if she was intruding and hung back. They parted and hovered there, looking at her mildly. Together they all surfaced.

Responses to people varied. With young males they were exuberant and speedy, and with older people they moved slowly and sedately, depending on our comportment — but not in every case. Some of the older divers could move very fast. Young Richard was a quiet, diffident person but they still bounced around him.

Mike tossed a yellow slate to the dolphins. As it sashayed down, they really enjoyed playing with it. For a time, as it lay on the sand, they cruised over it on their sides to inspect the dolphin drawn on it.

A group of five approached, including Mother and Remora Kid. The six-inch remora was active, at one time settling over the baby's eye. Mother seemed to restrain her offspring which was intent on play. As it zapped around Mike, she would get between and shepherd it away. From beneath her belly, Remora Kid peeked out and then swam around, avoiding its mother to engage with us.

At dusk the dolphins focussed around the speaker while I played them a variety of solo instrument recordings and we watched. There was nobody in the water. Then I entered with a 'waterphone' drum and the dolphins seethed around me, chattering wildly as I beat out a rhythm, dolphin-swimming in tempo. But it grew too dark to continue. A storm was approaching. Next day it rained, thundered and blew — the worst electrical storm I have ever experienced at sea. The Bermuda triangle *is* a peculiar area, weatherwise . . .

Day Five (22 July): Around 10am, four dolphins arrived so we matched them with four divers: a gentle, intimate session. Then Chopper, Stubby and several familiar dolphins joined them. Dane had a ferris wheel of dolphins

170

rotating vertically around him. As I swam and beat the drum, a frenzy of dolphins whistled and chattered around me. I could see nothing but dolphin beaks. Larry was huddled like a foetus on the surface, his knees drawn up, arms tight across his chest, changing 'I love you' to a cloud of dolphins as they brushed his arms, shoulders and mask. This was ecstacy for Larry.

Below him, in her own dolphin cloud, Jan stood on the sand totally enveloped. As she looked into their eyes, one brushed her shoulder gently. Mike had a group so close to his video camera he could only play. Later he explained that he was teasing them just as he would his little daughter — backing off as if to say 'you can't catch me', fingers and thumb close to his mouth like a beak opening and closing in a pseudo-threat, making chattering noises in his throat. Three dolphins, inches from his hand, were reciprocating, their jaws clacking, the sea buzzing with their chatter. As he retreated they advanced. They actually seemed to appreciate this mock alarm-and-pursuit game — an hilarious sight.

Eventually, there were 20 dolphins interacting with people on the surface, in mid-water and on the bottom. Whether scuba diving or snorkelling, it made no difference — everybody was intimately engaged with this group of familiars. Out on the fringe, Jan noticed another mother and baby. The mother had a series of fresh gashes near her tail. As Jan focussed on her, she faded into the blue haze. All was not sweetness and light out on the Banks.

At the day's end I went down on scuba, near the anchor rope. Jan snorkelled down and joined me and we were swimming together when Didi approached, alone but for her trailing remora and three rainbow runners alongside. What *is* up with Didi?

The Second Voyage.

Day One (2 August): With twelve people aboard, *Dolphin* dropped anchor on the Bank at 1.30pm and within minutes eight dolphins came sweeping in, to be joined by six divers, including Ruth Samuel. A newcomer, Ruth is a strong swimmer and diver — a slender, two-metre seaperson with long blonde hair. From the outset, she established a special physical rapport with the dolphins, whirling and spinning as they measured each other's capacities. But there were too many people in the water for the number of dolphins and this probably abridged the encounter — a problem at the beginning of each voyage as the decision to enter or wait had to be purely voluntary.

In the afternoon five dolphins approached, but to our amazement these were bottlenose (*Tursiops*). They had a more sedate manner, kept their distance and spent more time on the bottom. There were three adults, a juvenile and a baby. One adult had a ragged scar on its left pectoral. After 12 minutes they left, one making five distinct blows in quick succession, sending up spray each time. For Jan and me it was wonderful to see bottlenose again, so far from those we had met in New Zealand.

Twice that afternoon they returned briefly, in the company of spotted dolphins. On board in my sound studio, I was transmitting a recording of New Zealand bottlenose dolphins interacting with our musical output, a session we had had just before flying to Florida. Jan, Ruth and Richard entered the water where three spotted dolphins were close to the speaker. To my astonishment, each time the New Zealand dolphins made sounds, the Bahamans would respond in chorus. But during intervals of classical music they were utterly silent. Jan watched them hanging suspended below the speaker, as

171

if on an invisible thread, listening to the guitar notes: Rampal playing *Sekura*, Japanese seashore music. She lay there, not wanting to interrupt.

When Richard (18) dived, the dolphins swam around him in fast circles. Jan was not sure that she'd been noticed but as they hurtled by their eyes switched from Richard to her and they spun around her, one defaecating as it passed. Richard had noticed the same behaviour toward himself. A juvenile took a great interest in Richard, circling him at close range, scrutinising him intently.

Up at the bow, watched by those on board, Ruth was deeply involved with Mystery, an adult dolphin that had two remora, of different sizes. This dolphin had distinctive fin damage — definitely not Didi.

Of all the Maravilla dolphins only three were seen with remora. In Hawaii many of the spinner dolphins have them. Why only certain spotted dolphins carry them is a puzzle. There is nothing distasteful or parasitic about these sleek hitchhikers. Several times I have had them attach to my thighs. They have no more impact than a snowflake. I felt quite flattered. Their bodies are exquisitely sculptured to minimise drag and they resemble miniature sharks. They do not have a sucker pad, but an elliptical organ on top of the head rather like the slats of a venetian blind, hinged back to create a suction. As soon as the fish moves forward the slats fold down and release. I wish I could patent such a system for household use!

Ruth and Mystery would cruise along scrutinising each other. Mystery rubbed herself repeatedly in the sand and then nosed it, as if seeking food. As she rubbed her sides, tail and back on the bottom, Ruth imitated her. At this, Mystery slapped her tail on the surface, moving excitedly in a tight circle. Ruth copied with her swim fins.

Meanwhile, as the other three dolphins left, Richard decided to get out. He later explained to Jan that with three people in the water and only one dolphin, he thought it best to withdraw. Jan doubted the lone dolphin would have felt threatened by the presence of such a sensitive person.

Mystery now swam close to Jan, who wondered if she was crashing in on Ruth's game and withheld. Ruth gave her the 'OK' signal and told her later she felt she was introducing a friend to this new dolphin after spending time with it alone. Ruth was thinking, 'This is Jan from New Zealand. She's come a very long way and plays with dolphins often.' Such was their empathy, that Jan got the feeling she was welcome before the signal was made.

Now in the music field alongside the catamaran hull, Jan and Ruth had an intense session with Mystery; one of considerable importance to us all, watching or filming from above. The enjoyment was just as great as if we were participating. Had we all leapt in, nothing of this complexity might have evolved but by withholding we became richer in understanding. In fact, by being where we were, we could not fail to participate. As it turned out for most of us, this was to be the only opportunity of the voyage to swim with dolphins, but nobody regretted their forbearance.

Larry said it was as if there was some sort of electricity flowing among those in the water. Something extraordinary was certainly unfolding. From the deck Edee called to Jan, 'There's a piece of seaweed, if you need it.' Jan held the weed out to Mystery each time she passed by. This had been repeated so many times Jan was thinking, 'Maybe I'm too persistent,' and let the seaweed go. Just after the thought crossed her mind Mystery leapt joyfully and slipped away to return with her own scrap of weed, draped over a pectoral. She dived to the bottom and returned with the weed around a tail fluke. So deft was the changeover nobody saw it happen. (This is common and seems a part of the game.) Then Mystery gently trailed the weed over the sand, making pretty patterns as she went. She rose to the surface and approached Jan closely, looking directly into her eyes as if to ask, 'How did you like that?' Jan was bursting with joy and admiration. A pattern developed between Jan and Ruth where each seemed to know when it was time to dive down and frolic with the dolphin. Ruth felt she was being played with by the dolphin rather than the converse. Mystery swam along the bottom making patterns with her tail. Ruth dived and copied. The dolphin responded by wagging its tail at Ruth as it swept close by. In turn, each woman would dive down twisting, turning and undulating as the dolphin mimicked their every move. At one time Mystery approached Jan upside down. Jan turned over but her weight belt bumped clumsily on the sand, spoiling her rhythm. To Ruth's delight the dolphin emulated this clumsiness perfectly. For a dolphin to appear clumsy must demand an effort and it was most comical to watch. Mystery whacked her tail on the surface. The women did likewise — five times in succession, with the dolphin raising its head above the surface between times, cheered by those on deck. Clearly Mystery was surveying the audience. Later both women recalled that the music around them in the sea heightened their creativity — though whether it impressed the dolphin . . .

These games continued until 8.45pm. The weather had been deteriorating and the wind gauge now indicated a steady 25 knots. Only a catamaran could operate with any comfort in such a situation. The moment Jan began to tire, Mystery, with one final tail slap, left. For nearly two hours the trio had maintained communicative gamesplay of the highest order. As the women staggered aboard in the twilight it was hard to get them back into their bodies.

Later, Jan spoke with Janet Howe, diver and cook on the vessel under charter to Hardy Jones, which had anchored on the Bank for 25 weeks that season. Janet said they had seen this twin remora dolphin and named it Mystery because it always kept its distance. Furthermore, they had not seen Didi all season and thought she might be pregnant.

With serious mechanical problems wiping out our electrical system, and the onset of the storm, our second voyage had to be abandoned after just one memorable encounter. The remoteness of the Maravilla dolphins will often demand such a price.

The Third Voyage.

Day One (15 August): *Dolphin*, with 12 aboard, arrived on the Bank at 4.20pm, met by six dolphins. Nineteen-year-old Jim Gosling from Toronto sighted them first. Training as a biologist, Jim was eager to specialise in dolphins one day. He was so keyed up, donning his gear in a flash, that we said, 'Go on Jim. They're waiting for you.' And so it seemed; two adults and four juveniles right by the stern platform. Jim swam out with them around him. One adult came close. Jim lay still, looking at it. The dolphin did the same, gazing into his eyes. Jim said he felt like touching it; he only needed to poke out his tongue, but he didn't want to spoil the relationship. As they surfaced close to the hull, I called and one released a long stream of bubbles. Dean leapt in with a splash and a dolphin whacked its tail. Larry shouted that he had seen Remora Kid and its mother. I saw Rosemole and Scallopfin.

By the bow Ruth was soon involved with 13 dolphins including two bottlenose — Chopfin and companion. With her stamina and water skills, Ruth seemed especially attractive to dolphins. A group of eight wove tightly around her, often within centimetres so she could see nothing but dolphin bodies and dolphin eyes. Each time they came, she would hear their click trains loudly before sighting them. When a bottlenose got in front of adult spotted dolphins they snapped at it and made unusual 'Donald Duck' squawks and chatterings as if annoyed. One spotted dolphin carried seaweed on its tail while another tried to nibble it off. The weed-bearer kept it just beyond reach but eventually let the other take it on its pectoral. This time Ruth actually saw the exchange. Then it returned, the weed draped over its beak.

From the starboard bow Larry and I watched as an adult sped towards us on the surface, seaweed draped over the pectoral nearest us. As the weed slipped free, its tail flexed a little and consummately caught the weed on one fluke. This felt like a clear demonstration, answering the question we had been asking — how do they make these lightning changes?

When a new dolphin approached Ruth, they dived together and played the whirling game again and again, tighter and tighter until her fins almost touched her head, the dolphin almost against her stomach, but never in contact. She would focus on one dolphin and it would stay with her. A couple more would join in and then shift to another diver, the original remaining. More and more adult dolphins came in until Ruth estimated 28 at least.

One large adult broke off from the melee, and settled on the bottom near the bows, flukes and pectorals buried in the sand. It just lay there, motionless for four minutes. Ruth swam alongside it and I approached gently head-on, a little afraid something was wrong. It just eyed us, head nodding quietly, the base of its tail churning slightly in the sand. Dolphins approached at right angles, almost nudging it. I lay beside it for some time uncertain about whether it needed help. Eventually I accompanied its spiralling ascent. No hungry gasp for air, just a casual breath-taking. Later we realised Jack, the heavily built cameraman, had been doing *exactly* this just a short time before. Sprawled on his elbows, wearing scuba and no fins, inert, eye to viewfinder, Jack had been trying to capture the action.

During this time Larry had been dancing around the deck, leaping and shouting like a troop of monkeys. Edee saw a dolphin thrust its head out and make chattering sounds at him. As soon as Jan and I swam belly to belly a dolphin couple whirled around us in the same manner, whistling rapidly. I saw an adult upside down, its dorsal ploughing the sand. With my artificial dorsal I did likewise. At once it approached closely and rose beside me, eyeing me gently. For as long as I could keep it up, we dived and turned in unison. Other divers were enjoying similar relationships — it was an orgy of imitation, either species initiating a behaviour and the other responding. To me, in retrospect, all this is the primal level of communication — what we had shared with the children in Melanesian villages, gaining insights into each other's capacities without the use of language.

It had begun with six dolphins and two people as the rest of us on the bows watched, filmed and interacted with those in the water, offering insights to the divers gained from a broader view of the action, but, as more dolphins arrived, they were matched with humans entering the water.

Over towards Florida, the sun was nudging the horizon leaving the Gulf Stream and the Little Bahamas Bank to darkness. In the last glimmers the dolphins began to play with each other. An amazing sequence of patterned leaps: pairs vaulting in the same direction; then one belly up and the other normal, from opposite directions — the yin-yang leap we had seen in New Zealand. Down below I could just see their forms weaving in a seething mass of flexing bodies. Some were charging each other at great speed, then veering off, to rub along each other. There were pectoral caresses, copulatory pairings and rough, boisterous joustings. I saw two charge each other like goats, colliding blowhole to blowhole with an audible 'thunk' which I almost felt ten metres away.

As total darkness forced us out, the action seemed to be intensifying. Larry switched on powerful deck lights, rigged so they shone out over the ocean floor, dappling the sand prettily with ripples projected from the surface. At first the dolphins seemed a little startled and did not enter the perimeter of radiance. Edee saw a pair dodge away. But eventually they all came back and we rejoined them, mingling for a while in the artificial moonlight. But exhaustion had set in. This had been a monumental interlock. Jan and Edee and I sat on the heaving stern platform in the dark and hugged each other for joy, our legs too tired to resume their earthly load.

Since the catamaran's arrival, the dolphins had been with us for four hours. Had we all leapt in at the outset we could never have maintained such a level of activity. A girl told me later of coming out to the Maravilla dolphins in a dive boat. When 30 divers leapt in with one dolphin, it left immediately. No more were seen. . .

Day Two (16 August): Not surprisingly, after such a banquet, no dolphins approached the ship during the next day though we saw them leaping and feeding in the distance. Then, at 7.30 that evening five dolphins arrived, joined by five people in an intimate, quiet encounter.

At about midnight sleepers on deck became aware of dolphins feeding all around the ship in the moonlight. We could hear their breathings and chatterings, whistles, squawks and tailslaps over the calm sea. Deck lights were switched on and Jan spent an hour in the water but no dolphins came within a hundred metres of the ship. I think they were busy feeding. Several times squid came past and some sheltered under the hulls.

Day Three (17 August): Around midday eight dolphins, accompanied by 30 rainbow runners, approached. After eight minutes interaction they left. Activity was in progress in the vicinity where a lot of fish were herding.

That balmy afternoon there was a high haze in the sky and not a breath of wind. My diary notes: 'Bill and Ruth are half a kilometre from the ship, swimming over-arm in perfect unison. Dean and Jacqui are cruising in the zodiac. Larry is below us on scuba. Edee is joining up her freckles in the sun. Jan is in the bow hammock, the warm sea kissing her back. Toronto Jim has just made a Canadian salad and cooked spare ribs for dinner. Casimir is smiling and polishing brass in the rigging — "I must have been a metal smith in a past life". Mary is snoozing after a long car journey from Massachusetts and then straight to sea. Jack is chewing apple leather and I am pen pushing.'

Around 5pm that doldrum day, Edee decided to go for a swim. 'Would you mind putting on some music for me?' she asked. Just as she reached the bows, two dolphins met her, as if by arrangement. Nobody saw them coming. All on board gathered on the bows, spellbound as the most tender and subtle of encounters evolved. Edee is a quiet, unassuming person and it seemed appropriate that she should have her own session with the dolphins, uninterrupted by more boisterous companions. The young dolphins came so close that Edee would withdraw slightly to avoid collision — probably unnecessary but in keeping with her manner. She spoke to them continually and they eyed her very soul as all the crew looked on with delight. For quite some time nobody moved; no cameras clicked. We were all entranced at this very special interlock, right between the bows. Larry and I lay in the bow hammock only centimetres from Edee in her ecstasy. Then a third dolphin joined them; a larger juvenile that seemed to take charge. Perhaps it was a summons to resume the fish herding enterprise nearby and all three left together. Watching from the ship could be just as rewarding as being in the water.

Later that evening nine dolphins, accompanied by tuna, came in for 15 minutes interaction; with them was Chopfin, the bottlenose. At two in the morning dolphins swam close around our bows. Cas, in the deck hammock, didn't like to awaken the ship.

Day Four (18 August): All day there were dolphins in the vicinity, including Chopfin, but no interaction. We did not push it by bringing them in with the zodiac — a common practice on many trips to the Bank.

At eight in the evening, two adults with babies arrived at the bows. One was the tiniest we had seen, with not a blemish on its miniature body. It seemed incongruous but it was Jack's turn to enter first — great galumphing Jack! He left us with a resounding 'kerplosh' and fin-thrashed upcurrent like a baby whale. How would the dolphin mothers accept this? In his excitement Jack charged straight up to them. The dolphins dived. The adults circled Jack, checking him out, keeping their babies on the opposite side. Jack upended, bringing his legs out of the water and dived. Immediately a mother and baby mimicked his jack-knife dive perfectly, putting their tails high out of the water, and then pursued him. Below, the two babies came close to him, and surfaced by his side. We thought Jack would swallow his snorkel: it seemed the mothers decided he could be trusted. After six minutes they left. Jack was deeply moved and said he would probably never be the same again.

The expedition came to an end when mechanical problems plagued the catamaran. On the voyage home we had plenty of time to talk it over. Larry noted that we had not seen Didi at all this time. Ruth mentioned that on one dive her electronic watch had gone haywire. I recalled the diver on Hardy's boat who had used such a watch to signal the dolphins. Had Ruth's been ensonified? Jan told Larry how wonderful it was to meet dolphins with a crew dedicated to understanding their needs. She felt spontaneity was so important: 'We think we are playing with them, but it is the other way round. No way can we manipulate them or induce them to do what we want. When we mimic them they imitate the awkwardness of our mimicry. In Hardy's group, Janet Howe told us, the windup fish toy created an amazing response the first time but when they tried it again, nothing happened. Maybe it's because when we repeat anything we are in a different frame of mind. We're just a tiny bit bored with it and this lack of spontaneity is sensed by the dolphins. We are now experimenting with them whereas the first time it was truly a game, a novelty and we showed more joy and creativity.'

The Final Voyage.

Day One (5 September): To our delight, Mike Daniel decided to join us with his 16-metre aluminium yacht *Tara*, a fine ocean racer bought with silver from *Maravilla*. This would be her maiden voyage since having a complete refit. With him came his wife Nancy, their four-year-old daughter Michelle, and two crewmen. Michelle was obsessed with the wish to swim with dolphins. I had long been curious as to how they would respond to a small human. This would be a special trip, having two vessels — a small ocean community.

Credit: Dan Sammis

special times in 1985 the spotted dolphins of the Bahamas gave patterned performances and would accept sensitive
mans such as Ro Lotufo and Carol Ball in their 'dance'.

PLATE 26A   Credit: Dan Sammis PLATE 26B   Credit: Dan Samm

The dolphins often mimic the vertical stance of divers, poised motionless in mid water, and even falling backwards o
balance.

PLATE 26C   Credit: Dan Sam

The circle ritual is a dance-like manoeuvre, performed while three divers watch in awe. Many of the dolphins are recognisa
individuals such as *Scallopfin* (or *Chopper*) and *Stubby*.

Credit: Dan Sammis

th divers who visit them frequently, such as Ro Lotufo, the Bahamas dolphins have complex and sensitive relationships
ereas casual visitors to the area may not experience such profound and extended interlocks.

Credit: Dan Sammis

PLATE 27C

Credit: Dan Sammis

PLATE 28

Credit: Dan Sammis

A rare and supreme moment in human/dolphin interaction as *Romeo*, from his social group, approaches Ro Lotufo to be caressed. Languid and passive in human hands, he falls tail first to rise and repeat when beckoned.

With people moving to and fro there would be something new to interest the dolphins.

Another pleasure would be a dive on *Maravilla*. We had met Bob Marx back in port and he was agreeable that we visit the wreck. Their treasure diving vessel, *Rio Grande*, had just been working the site again. The crew told us that when they arrived, dolphins visited them for 45 minutes. They had had another brief encounter but after that, although they saw them in the vicinity, the dolphins stayed out.

With 11 aboard we anchored on the Bank at 4.15pm, and were met by seven dolphins, including Scallopfin, Rosemole and Chopper. Mike tossed out his snorkel in his usual greeting and then yelled, 'One of the dolphins has just crapped on my snorkel.' Perhaps it was an acknowledgement. He saw a dolphin pluck a piece of seagrass from the bottom and mouth it playfully. It passed Mike, pushing the morsel towards him. Then it swam up to him and ate the eel grass as it regarded him steadily. At the bow of *Tara*, a mother and baby approached Mike. The mother left her baby beside him, circled the yacht and resumed her charge. Those on board were impressed with her trust.

I met Rosemole and Doublenick lying on the bottom, scarcely moving. We played fast swimming games, rubbing our backs on the sand. Then one released a big gulp of air. When I did likewise it responded enthusiastically.

By 7.30pm more and more dolphins were arriving: Remora Kid and many familiar to us — even one adult whom I suspect may have been Didi. It was so dark we could barely see but when I called 'Di-di', the dolphin responded as before, swimming in an arc around me. Near the bottom, a group of ten adults were twisting and turning in a huddle, making an incredible array of sound, whistles in varying pitches and sonar clicks with an electric intensity we could feel all around. We were getting too far from the catamaran. Bill whistled us in. As we returned over the black ocean the dolphins accompanied us in a tight bunch, right up to the stern platform, then left.

Day Two (6 September): The day started with dolphin visits to both vessels. Larry and I met five juveniles off the bows. I dived with a dolphin just above my back. From the deck Will saw it defaecate on me. From the context and from other occasions, we would *like* to think this is a greeting. Unaware of the honour, I carried on swimming, entranced by Handel's Water Music.

The morning ocean was sparkling clear, warm and only six metres deep. For once there was no current. I became intrigued with the discreet community of marine life, the sand bank's best-kept secret. Tiny flounders danced upright, mouth to mouth in fluttering courtship. A minute wrasse, like a gem with an emerald spot on its side, hovered close to a sponge growing from the back of a half-buried conch. Portulid crabs, scurrying over the bottom, would stop, hunker down and burrow until their eyes were concealed, their bodies still visible. I noticed a weird form, a pear-shaped balloon, gliding towards me. As I went closer to investigate, it resolved into a large barracuda stalking me. Time and again it would zero in, its flanks cunningly concealed by this head-on approach. The jaws were quite invisible. By some masterpiece of colour and form, the predator could advance on prey to close range and then make a lightning lunge. A second 'cuda joined it. When one veered at right angles and made a threatening jaw gape, I replied, parting my fingers and growling into my snorkel. This body language seemed effective: it turned and fled but its companion hovered nearby. The tiny carangid fish, riding the pressure wave of my mask, saw it two metres from me and decided to switch 'horses'. As it approached the 'cuda from astern, the big fish began threshing its tail and twitching with annoyance. Clearly it recognised this gadfly's intentions and didn't appreciate such company. Maybe it gave the show away when the 'cuda stalked prey.

One of the things that was concerning me was our poor ability to recognise dolphin sexes clearly. We had discussed this in the saloon, hunting through reference books. Many diagrams are misleading as they don't include the navel when showing the pattern of anal/genital apertures. Lyall Watson's book *Whales of the World* solved the problem with a very clear diagram and description (p. 22). Now I wanted to test the picture I had memorised.

Suddenly, I heard a whistle as I rose for air. Was it my sinuses squeaking? More squeaks and whistles. A young dolphin hove into view. To meet a cetacean while free-diving alone in mid-ocean is more like a dream than a reality. What gender is it? Body language for this gesture is obvious. Our culture censures the gesture and in the company of others the idea had not arisen. I swam nude as the dolphin circled. It came in and rose in front of me. At close range the pattern of orifices was perfectly clear — I was meeting a young female. We circled several times. Then, on the edge of vision, I glimpsed five more dolphins just above the sand, an adult and four juveniles. The young female joined them. I headed back to the ship to see if an interlock would ensue. Nobody on either vessel had seen any dolphins as they were remaining close to the bottom. This encounter would not have been so memorable were it not for the events that followed that afternoon.

For the rest of the morning the dolphins were busy feeding nearby. A school of silvery sprats took refuge beneath the ship. We had several brief visits and the little girl had an initial swim with dolphins. Her mother was amazed at Michele's eagerness to continue and there were tears and protests when it was time to get out. Conditions were not ideal.

Around midday *Tara* set out along the Bank to visit the treasure ship, Mike conning us from the mast, his wife Nancy at the wheel. I was surprised how close the wreck was. Only a kilometre from where the catamaran swung at anchor, Mike said, 'This looks like it. C'mon Wade, let's see if we can find the markers — a car tyre, an airconditioner and junk.'

I dived and drank deeply — a pretty scene. Sixteen metres below were thickets of purple staghorn coral fringing a sandy basin patched with eel grass beds, sponges, sea fans and myriads of reef fishes. In all our time on the Bank, I had never suspected anything more than the vast, featureless sand plains. I never dreamed of such an oasis of reef world — a rectangular block of coral. Could it be man-made? It was coral. Then a cylindrical object — a diesel filter. Nearby, a beer bottle; then, on a coral head, two concrete blocks linked with braided nylon; another further on. Beneath the sand, perhaps six metres below, were the hidden relics of human tragedy. Here, on a mid-ocean sand mound, 600 people had met disaster when trying to round the Bahamas too soon; eager to reach Spain with their opulent plunder from the Americas.

I felt a challenge to free-dive further than usual: 20 metres plus. The hand of pressure feels gentle and reassuring as I hedge-hop over the bottom. With so much to delight the eye it seems easier to remain below. Not a fragment of shipwreck on the coral screes. Over the white sand no glint of gold or the black silver discs I knew from my own treasure-diving days.

Mike explained later that all the ship material is deep beneath the sand in the white avenues and basins amongst the coral. Only the powerful downdraft of the treasure salvor's propellers could reveal any human artifact. Cyclonic storms from the north-east would churn and seethe over this Bank with all the fury of an enraged Atlantic. The tranquillity I was enjoying could not persist for long in the hurricane season. Time and again the salvors had blasted huge craters down to the wreck and as many times, the ocean obliterated their work. Mike could see traces of where Bob Marx had prospected with a magnetometer only the previous week. But Mike could find no sign of his pet grouper. Topside he explained that in all the years they had worked the wreck only at one phase had salvors dropped litter on the site. They soon learnt they only got their own back. Somewhere beneath those shimmering sands still lay the treasure of treasures — a solid gold life-size statue of the Madonna.*

Treasure enough for me was this chance to gain a mental picture of the place where men and dolphins have established the deepest rapport on the planet. Before returning to the catamaran we collected some lobster, conch and fish for a twin ship banquet. Would the dolphins sense the community and friendship between the vessels?

As we circled the catamaran, Jan and Larry were fizzing with excitement. 'The dolphins came,' Jan yelled. I was to learn that my nude display may have provoked a rather special demonstration and there was a suggestion that the dolphins were able to communicate about us to each other.

That afternoon, skipper Bill Griffin told me, he had had a gutsful of ship maintenance. Down in the bilges the macerator pumps had burnt out; hence the stench as all the shower water backed up. Bill decided to take a spell, curling up with a book in the cockpit. Then 'pssht', blowhole sounds and five dolphins were alongside. But everyone was siestering below. He peeps down — just as Jan awakes. She went up but decided to let Bill establish a relationship alone. For 15 minutes she watched and wrote notes: 'They were right there to meet him as he entered the water. As he swam out they were close around him.'

'I was just about breathing them,' said Bill afterwards.

Later, when a dolphin thrust its head out at Jan and she recognised Spot, she decided to get in too. For the next half-hour they had the most special of sessions. These dolphins were moving so close and so slowly, they were almost stationary at times. It was incredible that they could move at all and yet not touch the humans. *Every one of the five was a female*; there was not the slightest doubt. They were all young adults with just a few spots.

'It was as if they *wanted* us to know they were females,' said Jan. 'They kept coming close and presenting their bellies at eye level, gliding slowly alongside us. As I rose to the surface one was four inches away, its anal/ genital and mammary slits directly before my eyes. I could have put my finger in it. I wanted so badly to stroke one, just a gentle touch. But no — I couldn't risk spoiling their trust. Dolphins have come close before but not in this slow, deliberate and sensuous manner. One swam belly to belly with Bill several times . . .'

Down on the sand the five females bunched together, two belly to belly, the top one stroking its pectorals to and fro on the genital zone of the one beneath. This was repeated frequently. Another pair swam with their fins touching and rubbing — so Bill and Jan swam with their hands clasped.

* About 600 kilograms weight of gold.

The dolphins tail-walked on the sand, rubbed their backs on it, swam upside down and barrel rolled as the humans strove to emulate them.

Then there were games with seaweed. Instead of grasping it, Jan scooped it on her arm as dolphins do with their pectorals. At that, several dolphins started playing with seaweed. One dashed in, scooped up a huge piece on her tail and dived with it, releasing, catching it on her beak, then her pectoral. By this time Larry was on deck video-filming the encounter. The dolphin thrust her head out and flicked the weed to Larry over the surface. Larry filmed. Then a dolphin released a bubble-gulp. Larry recorded this too. He recalled my conversation that morning about the significance of bubble-gulps. I had seen many dolphins make this gesture over the years and came to suspect it was important. I showed them a passage in Karen Pryor's *Lads Before the Wind* about captive dolphins blowing bubbles as a positive, affirmative signal. I described my own bubble-gulp experience of the previous day and said I felt the meaning depended on the context.

Larry had a premonition that this dolphin was going to do something special. Nicholas was standing beside him. Larry was calling the dolphin, talking to it while his camera was trained on it. Just beneath the surface alongside the hull, the dolphin released this huge bubble-gulp and then swam around exuberantly. Its rapid movements contrasted clearly with the slow, deliberate swimming of the other four, still involved with Jan and Bill. Larry was sure the bubble-gulp was a positive communication: eye contact, sound, the bubble explosion and then excited swimming.

The session with Jan and Bill lasted 40 minutes before Cas entered with a video camera. By this time more dolphins had arrived: 12 in all, including several adults. The current had increased, reducing clarity and Jan was feeling tired. The mood had altered and the dolphins no longer came so close but they really performed for the camera. They repeated all the seaweed stunts and Cas got an unforgettable sequence where a dolphin approached a piece on the bottom. Standing erect beside it, with a tail flick it whisks it up and catches it on a fin, much as a skilled soccer player can get a ball into the air with his instep. All their antics were repeated for the camera: belly to belly swimming, tail walking, upside down swimming. (Viewing the film, it is hard to believe it is all a repeat performance.) Eventually the current was too strong and the humans had to leave.

Tara returned from the treasure ship and anchored 300 metres from *Dolphin*. As the zodiac came over to collect the seafood for dinner, six dolphins appeared.

I got into the water and was surrounded by the adult White Blaze and five juveniles, including Remora Kid. I have never had a more intimate encounter with the dolphins mimicking my every acrobatic turn so near I had only to make the slightest move and we would have touched. But I didn't. That, it seems, is an essential part of the game. White Blaze came close, too. She seemed like a kindergarten teacher with all her charges pirouetting around me, skimming the sand or gliding slowly by for eye contact. I was just below the surface when a small dolphin swam over my shoulder from behind. A metre in front of me, viewed from the tail, a worm emerged from its genital slit, withdrew, emerged, wriggled — a young male displaying its erection. I felt a touch on my arm as I broke the surface. A tiny hand greeted me. Michelle and Nancy had swum out. Nancy said that when I got in nothing would hold Michelle. On deck we had been playing with her felt toy dolphin, Dolly, tossing it to and fro, spinning and flipping it as we fantasised dolphin games. And now it was really happening. The dolphins swam so tight around our human trio that I could see nothing but dolphin bodies. Michelle and Nancy would dive down together and look up into a maelstrom. Not the least flicker of fear in the child. I found this amazing that she should fearlessly enter the sea with no land in sight. At one stage from *Tara*'s stern Mike saw a dolphin pass between his daughter and me. I could scarcely believe that possible because the child was right beside me at the surface. I was concerned for her safety, she was such a tiny human in all that immensity. Before this day, Michelle could swim but had never dived below the surface. The moment she saw the dolphins she just followed them down and when it came time for her to leave she tried to evade her mother by swimming away underwater. She wept as she climbed the ladder and for the rest of the day kept stroking her shoulder and breathing, 'a dolphin touched me'. I told Mike about the young male's display as his daughter entered the water. 'I'll kill him, I'll kill him,' shouted Mike in mock horror, while his daughter purred, 'a dolphin touched me.'

Bahamas Epilogue — Following our return to New Zealand, Jan and I kept in touch with the Maravilla dolphins through friends in the United States. Mike sent us a video and photos of Michelle with the dolphins. One shows her rising in tandem with a juvenile, her tiny hand outstretched towards its underside. In the picture the surface is not visible.

Larry continued making voyages, season after season (1985-87). With frequent visits from the same sensitive people, interspecies relationships developed to the point where *touch* was as intimate and complex as in any of our solitary dolphin episodes. One of the most important lessons to be gained from the Maravilla dolphins is the rule, 'Don't touch until invited'.

For many people, physical contact with a dolphin has great importance. We are accustomed to stroking and fondling domestic animals and, without reflecting, we may try to extend this to wild creatures. When a woman on one of Larry's expeditions was tempted to touch one of the quartet interacting with her, it took off rapidly, leapt out and all four departed for half an hour. Without prior communication and familiarity, touch can be seen as implicit dominance behaviour: touch, grasp, ride, harness, control.

It seems that to dolphins, as with most mammals, touch is a highly significant stage in a relationship. Mutual trust and respect must be established first. As among humans, it usually requires repeated acquaintance and empathy before physical contact is accepted. Most often the dolphins initiate it. With our manipulative, grasping limbs and history of aggression, we are the dangerous ones.

As part of the crew on *Dolphin*, Dan Sammis and his lady Ro Lotufo had the opportunity for frequent and sensitive meetings week after week during the 1985 season, including some quite elaborate acoustic exchanges. In May, Dan had a strange encounter with six adult dolphins that tail-stood in a close, touching circle, their heads pointed up, regarding Dan as he tried to explain his body to them. They sonared him heavily as he showed his hands and how he could move them. A commotion ensued and then a female advanced the intervening two metres, gently touching his mask with her beak and passed through his arms. She rejoined the group and they all ascended for air.

After a subsequent trip our friend Ruth wrote: 'Three dolphins came and stayed for an hour: two adults with one baby. They played and played. So did Ro, Danny and I. The large male came over to one side with us. Danny tentatively stroked the dolphin's side with the back of his hand, and then his palm. Ro briefly stroked him. The dolphin came to me, nose-to-nose, so close I was cross-eyed. Then Ro came over to us. The dolphin turned and Ro stroked his side. Lying on the surface, she was perpendicular to the vertical dolphin. With eyes closed, it slowly sank out of her reach, as she stayed on the surface. She gestured with her hands for him to come back up to her, and *he would swim up into her arms*. She would stroke him, and he would slowly slip down out of her reach. Again she gestured, and again he swam up to her. She gave him a gentle squeeze. She was very excited, but did not want to scare him. It was unbelievable to watch. The dolphin's eyes were closed for the most part, but would slightly open. He seemed so serene and relaxed.'

Larry developed a new game with two mothers and two babies. As he swam between them they folded around him in a languorous, mellow mood, eyes wide open centimetres from his mask. Then he would roll and go beneath them, rolling up the other side.

Watched by six people and nine dolphins, Sweetie came to Bill and lay on him while he gently touched her under the pectorals. She kissed his mask and touched both him and Ro.

The seaweed games became more complex. In 1984, Jan and Ruth had demonstrated a game as the dolphins watched, then the dolphins got their own bit of weed and showed manipulative skills.

A year later Ruth wrote: 'I pick up and swim with a large strand of seaweed. Two dolphins follow. As they near me I release the strand. One catches it in his mouth, spins and slings it to and fro. Finally he lets it go and catches it on his tail. He circles and swims beneath me, still holding the weed. As I dive after him he releases it for me to catch. He returns to me. Now what to do? I swim ahead and let the seaweed go. He catches and drops it. I pick it up — but he has gone over to the other divers. *We were really playing together*. Now three are back. I dive and twirl. They twirl and dive. The trio swim beneath me. Two are mating, intimately locked together for five seconds. The third is male and his penis is extended.'

With successive voyages new key dolphins emerged such as Romeo and Boyfriend, Scratches, Pictures and Nippy, Ladyfriend and Sweetie. From May 1985, Didi was seen no more, but Rosemole, Scallopfin and Chopper were intensively involved.

Fresh insights were gained. When a menstruating woman entered the water, an adult male became highly aroused, displaying an erection and leaping right over her. This same person, exhausted after lengthy gamesplay with another group of dolphins, said, 'I'm tired, I'm tired.' At that moment one dolphin released a large bubble-gulp and they all left.

On another occasion, Larry's son Dean entered the water with an injured leg and was stricken by cramp which made the upcurrent swim back difficult. A female accompanied him at close range all the way to the stern platform, putting her head out while she stayed alongside. When he climbed out, she left, swimming away belly up.

Bottom-feeding dolphins were seen stirring up the sand, flushing out flounder. Their retinue of jacks swooped in and snapped them up — another reason for following dolphins. Gradually the bottlenose dolphins came closer and began to interact with divers.

In May 1986, as three spotted dolphins swam parallel with Richard Carey, a large bottlenose came within inches of his face: 'With it shadowing my movements I spiralled to the bottom and back, DNA fashion, three times.

The dolphin looked into my eyes as though windows into my soul. It was a mutual acknowledgement of conscious awareness.'

Larry continued my acoustic communication efforts with improved gear. As physical encounters intensified, so did the quality of sound exchanges. With a hydrophone attached to the video camera, Dan was able to film a tête à tête wherein the dolphin made loud, distinct whistles to him at much lower frequencies than normal; easier for a diver to appreciate. It seems that with more familiarity, the dolphins began to adjust to human hearing capacities. In response to people talking to them they modified their acoustic output with low, smooth 'ooing' whistles, quite distinct from their usual signature whistles.

On three successive nights in September 1985, Larry had dolphins close around the hydrophone for extended periods from 10pm to midnight. Those on deck talked to them through a microphone linked to the underwater speaker and heard their amplified responses. The dolphins were frolicking right alongside, rubbing and touching before tail-slapping on departure. The sounds recorded are unlike *anything* I have ever heard dolphins produce. Eventually I was to replay them to a New Zealand dolphin.

During 1986-87, Larry's *Friends of the Sea* expeditions to the Maravilla dolphins were included in the schedule of the Oceanic Society, San Francisco, which advertises to its members a weekly timetable of expeditions that make the friendliest dolphins in the world accessible to those who are really determined. Few have been disappointed.

As Ruth Samuels wrote, the hardest part is leaving. 'Just the five of us, me and four dolphins cruising up current. They are giving me just enough room so that I can swim. I possibly have three or four inches clearance as I reach out my hands to swim, as well as at the sides of my body. Their eyes, or at least one eye, were on me all the time. At first I thought they were just sensing my presence, and managing to avoid a collision; then I looked into the eye of a dolphin directly ahead of me. They could see me very well, and were just barely giving me space to swim! I had been concerned that I would inadvertently hit one with an arm because I was swimming with both arms and legs rapidly moving, but they gave no sign of fear or concern with all my splashing as I swam. They don't even look like they are moving their tails or fins. They are just coasting. I am swimming as hard as I possibly can. One dives away. I am euphoric, and want to buzz the boat. We all turn in unison and blast back to the boat on a wave. I drop behind as I can't quite keep up. We pick up four more dolphins. I am now one of eight! They are older and more spotted. There are people and dolphins everywhere! It is quite confusing again. I can hardly breathe from the exertion, and my muscles are screaming for rest. I am totally exhausted. I decide that I have done it, and that I need to get out of the water. Everyone else is cavorting with the dolphins. On deck I watch the action and play, and my eyes fill with tears.

'I was one with the dolphins, one with the universe. I had my moment of joy with them. Such a rush of feeling, such sheer pleasure. The dolphins are so tolerant of us. Why do they wait and play? Why do they watch us awkward beings? Are we the funniest show in town? The interlock lasts approximately one hour and fifteen minutes. People get out bubbling and exhausted. We are now sailing back to Westend, as it is time to head home. The sails are filled with a strong breeze. It is so peaceful and quiet, and a little sad to be going back.'

Formal Study — Cetacean biologist, Denise Herzing, accompanied Oceanic Society expeditions to the Maravilla dolphins during 1985, '86 and '87 seasons. She had previously made a study of migratory grey whales in Oregon. Her approach is broad and full of insight, and will lead to deeper understanding. Asked for comment, her words provide a fitting conclusion here: 'I think one of the most interesting phenomena that I've observed with the spotted dolphins of the Bahamas is the gesture of extending their communication signals to humans.

'One morning seven dolphins came to our boat at anchor. There were four very old animals, two of which we had previously identified and sexed as males. Accompanying them were three adult/infant pairs, most likely mothers with their babies. The older males seemed to take the role of scouts, checking us out with their echolocation and keeping a distance between us and the mother/infant pairs. This interaction continued for 30 minutes until the dolphins left. I got out of the water and was on the stern describing the interaction that had just transpired when I heard a slapping sound on the water. Just off the stern of the boat was a dolphin tail-slapping. When I entered the water I saw the *same four males*, this time with five mother/baby pairs. We then were able to interact closely with all the pairs.

'Now tail-slapping, like other signals, can be interpreted in different ways, depending on context. In captivity, tail-slaps have been interpreted as possibly aggressive warnings, but also as an attention-getting mechanism to trainers. That one species is able to extend part of its communication system to another species is certainly not a new concept, but it speaks of possible channels to pursue in the field of interspecies communication.

'Interspecies research has been plagued by a species-centric mentality of teaching English to other forms of life. This is, ironically, the same thought process that early European settlers used when discovering native peoples — teach them English and enculturate them. This attitude leads to many faulty conclusions about measures of intelligence and communication ability in other cultures as well as in other species.

'Another interesting pattern of interaction stems from the bottlenose dolphins in the area. They do not interact with humans in any regular way or in close proximity, but I have observed them hanging out on the periphery when we are interacting with the spotted dolphins. I have recorded, visually and acoustically, bottlenose cruising the sand banks in a "spread" formation hunting for food, and these "searching" sounds are a distinct combination of echolocation squawks and bleats.

'The first season (1985) I was out as a naturalist, we would monitor these sounds at night. One night we tried vocally interacting with them using a microphone and underwater speaker. This exchange seemed, initially, like a dialogue, back and forth between dolphins and humans. Now, squawks are inherently the type of sound that can resemble human mimicry. Although, upon initial interaction this seemed to be what was happening, later analysis of the tapes shows that these dolphins, while retaining their own sounds, were mimicking what one called "prosodic" features of speech, ie rhythm, intonation, etc.

'These prosodic features of dolphin communication have only recently been explored scientifically and it appears that the space or time *before* the sound has a consistent relationship to the subsequent sound produced. This feature, in human language, signals the beginning or end of a statement, creating information for the second participant.

'The many gestural and mimetic aspects of communication systems offer a productive direction for interspecies communication research. Although other vocal mimicry has been reported (Lilly, Doak, Nollman, Ridgeway), other modalities may be fruitful avenues to explore (gestural, postural, tactile), hopefully leading to a better understanding of interspecies interaction.'

The Spinners of Brazil

Three hundred and twenty kilometres east of Brazil lies Fernando de Noronha, a tropical archipelago called by Brazilians the *Emerald of the Atlantic*. Volcanic in origin it consists of six main islands and 14 rock piles dominated by a rugged, 400-metre mountain peak. Scattered among the green hills are ruins of 21 forts built by the Portuguese who claimed the islands in 1502. Today it is populated by 1200 people, most of whom work for the military air base. A few are fishermen, tend small holdings or raise cattle.

As if to prove the friendly Maravilla dolphins are not just an oddity of nature, a parallel has emerged at Fernando de Noronha. The circumstances of the spinner dolphin colony in the Bay of Dolphins exactly fulfil criteria for a human/dolphin interface: resident dolphins with a handy food supply, remote from human population pressures, yet accessible daily to people wishing to establish familiarity and trust. Situations like this are rare so it is especially exciting that it has recurred with a different species; the spinners, *Stenella longirostris*, are among the shyest of dolphins.

We have on file reports of moderately friendly spinners in Guam and at Midway Island. In 1980, I swam with a colony at Lanai in Hawaii and saw another at Kealakekua on the Big Isle. Spinners are slender, long-beaked dolphins about the size of the common dolphin, but famous for their superb rotating leaps. They are nocturnal feeders, preying on squid in deep water and returning to certain sheltered bays to pass the day in courtship and socialising.

For several hours in the middle of the day, spinners have a rest period when they are silent, swimming together in large groups which breathe and dive in unison, maintaining group vigilance while, it is believed, they rest the acoustic part of their brains.

Whenever human populations are close to spinner colonies, friendly relationships may be spoilt by unwitting encroachment on the dolphins' rest period. From a cliff top at Lanai I watched two divers attempting to engage the dolphins. Time and again, just before the spinners got within visual range, they would dive to reappear some distance away. It looked like deliberate avoidance. Lanai is a popular tourist area, yet close relationships with divers have not evolved there, despite the proximity of the dolphins.

I first heard of the friendly Brazilian spinners in a letter I received while in Florida. It was from our Swedish friend, Lars Löfgren. Exploring Brazil for cetacean contacts, Lars found a village where bottlenose dolphins assist fishermen by rounding up mullet for their thrownets.* Then Lars heard of a military base on a remote offshore island where there was this fabled Bay of Dolphins.

Early in 1985, I heard from Janet Nowak, a diving guide at St Thomas in the US Virgins. She told us she had read my book and shared our interest.

---

* Cousteau filmed a similar situation on the opposite side of the Atlantic where dolphins assist the Imragen fishermen of Mauritania.

In September she would be going to Fernando de Noronha, with diving entrepreneur Russel Coffin, to spend time with the dolphins. A correspondence developed and in due course Janet wrote of the expedition:

'We made daily trips to the bay where we were escorted to our mooring by leaping, splashing dolphins. We had the bay to ourselves and of the six members of our party, I was the only one not burdened with cameras. Using snorkel gear I soon became surrounded by chattering, squeaking, curious dolphins. Cautious at first, we became good friends after a few days. I learned to play their games, but only by their rules. We would swim together, diving down, spiralling up and circling. I would be swimming with a large group when all of a sudden they would start to play, leaping, diving, defaecating, bumping into me and swimming away as if to say "hurry up". Eventually mothers would let their young approach and they were the most fun of all. Curious males would rub against me as if I were one of them. They were always loving, touching, mating, playing and teaching, as if they wanted to show us the way. I can still feel their "sonic wash", we became so deeply attached. Through the wonderful pictures taken Russel is hoping to persuade the Brazilian Government to create a Marine Reserve there.'

During this, and subsequent expeditions led by Russel Coffin, the friendliness of the spinners was documented with stills and movie in order to seek their protection. Russel was aware of plans for a major tourist development on these seldom visited shores and knew that it would take an international outcry to secure the dolphins' future.

In response to expedition divers, the spinner dolphins showed a similar interest in seaweed games to the Bahamans,* but when the human voice was broadcast underwater they all left in panic. Gradually underwater music was accepted but no positive response was noted. Their initial reaction to scuba was also instant flight.

A year later, Janet Nowak and Russel Coffin flew out to Fernando de Noronha, a 40-minute flight from Recife. Their first sight was the 20 kilometres of jagged coastline with the peak of Morro de Peco rising out of a green hillside.

Next morning, as their dive boat *Thor* left port, a trio of spotted dolphins met them and bow rode until they reached the Bay of Dolphins. As they entered, the spotteds swam out to sea. At this moment a group of spinners greeted the boat, riding the bow up to the permanent mooring.

Janet Nowak's Diary.

7 September 1986: I entered the water alone. As if 365 days had never happened, I swam to the dolphins, arms at my side. We picked up just where we had left off last year. I joined a large group of about 50. Mothers and tiny babies came so close to me I felt very welcome. A big male with a heavily scarred tail stock came in and watched me so closely, swimming ahead and returning several times. I named him Lumpy because of the scar. (This dolphin was important in later interlocks.) For some time we all swam together, diving down and spiralling up as before. To feel their sonar and hear all the playful chatter made me feel at home. I returned to the boat and thanked the others for allowing me a special time with the dolphins. Now all the group entered but when two people donned scuba the dolphins' voices rose in pitch and

* Tatters, a spinner in Hawaii, played with a plastic bag on his pectoral fin.

became confused. Grouping together they all left the bay — a mystery because a year ago they accepted the film-making team with tow vehicles and scuba. (Cameraman Jack McKenney records that the dolphins only accepted the scuba when used in conjunction with the tow vehicle: 'It was evident they enjoyed the motor noise but not the scuba exhaust.')

8 September: When *Thor* motored into the bay, the dolphins greeted us with joyous leaping and splashing. All was forgiven. I entered and swam off with a group of 40. My every move was imitated and improved upon. As I tired, the dolphins slowed their pace to await my next wind. Back at the boat, Russel was filming a manta ray using the tow vehicle and scuba. The dolphins showed no concern. Then I heard a strange sound, like an undersea owl — a medium-pitched 'ooo' repeated at intervals. I realised it was the minke whale and calf we had sighted earlier, somewhere beyond vision.

9 September: We arrived in the bay around noon after scuba diving on a pinnacle. There must have been 300 dolphins today. I entered alone wearing my silver suit. Instant interlock. Several of the larger dolphins scrutinised me closely. When Russel entered with the tow vehicle and scuba, the dolphins left at once. Why is it they enjoy playing with divers and machines at one time and seem spooked by them another? (It could be that they were not acceptable during the middle of the day, rest period, or that, as in Hawaii, different dolphins visit the bay from day to day.)

12 September: The previous two days were involved with wreck diving but we did sight spinners in the vicinity, including the easily recognisable Bentfin. Again the dolphins escorted *Thor* into their bay. I entered and became involved with a group of seven for the next hour. They were all so perfect I had difficulty in recognising any of this set, but then two more joined us, one with a shredded dorsal and the other with a cross-shaped scar near the tail. This pair came in repeatedly. The interlock was peaceful. I blended with the group and lost track of time — until my mask was dislodged by a pectoral fin as I was caught up in a session of rough horsing. It was as though I were invisible.

13 September: On our way, *Thor* was joined by both spotted and spinner dolphins. As we approached the bay the spotteds swam out to sea and the spinners entered. I had a sore throat and felt feverish as I got in. My energy was low. The dolphins reacted accordingly, their mood was tranquil and gentle. Quietly they passed above and below me in large groups. Their sounds were minimal. I became aware of large groups beneath me in 20 metres of water with me at the centre. I felt as though I were a participant in a group meditation. I closed my eyes and slipped into alpha, feeling weightless and safe with these friends all around. I floated quietly for an indefinite period. I was aroused by a high-pitched whistle and a loud slap. A juvenile had made a leap to enter right beside me. I now felt rested and exhilarated, and decided to warm up on deck. At 3pm as usual, all the dolphins went leaping and spinning out to sea for their night's feeding. Some rode with us all the way to port.

16 September: For the past two days we have explored the island, climbing the highest peak and hiking to the cliffs which overlook the Bay of Dolphins. Just before sunset, we saw them all swim slowly out to sea — a full moon night.

Today *Thor* set out at 8.30am to be met by seven spotted dolphins. Just after they left our bows, we rubbed our eyes in disbelief — there were five bottlenose in their place — the biggest dolphins I have ever seen, more like small whales, about four metres long.

In the bay I swam out to the spinner dolphins. They were quiet, making hardly a sound as I approached. I was quiet, too. A small group came over and slowly escorted me to the boat before rejoining the main group. I felt it was over for the day. Then my favourites approached — Bentfin, Lumpy, Scarface and Rusty. I rolled over to float on my back and the chatter began. They wanted to play — swimming in fast circles, leaping and spinning. I was mesmerised. We played long and hard. Without knowing it, I free-dived beyond my usual limits. Bentfin swam so close we bumped at times. More and more dolphins joined us and the play got crazier. At one point there were so many I could not see out of the pack. A male circled me faster and closer, and then invited me to descend. As I dived and spun he circled me like lightning. At this point *Thor* came over. The dolphins took up a vee formation on the bow and to my surprise, I found myself riding the pressure wave with them for a time. Suddenly a group of seven circled me, took up vertical postures and waved their pectorals. Puzzled at this behaviour I stayed still, arms at my side. I could not advance or retreat. I was captive. Next instant, I was being bumped and invited to swim. At that moment I could have touched any member of the group but didn't want to betray their trust. Russel entered with his video strapped to the tow vehicle and gave me a machine, too. As I jetted along at high speed the dolphins went berserk. We were well out of the bay by now and Russel dived deep with a large group just in front of him. When we got back to the boat I was so elated it was hard to believe I had been in the water almost five hours.

17 September: Today I swam off with a large group and was melting into their weightless world when something brought me crashing back to an awful reality. Swimming beside me was a large spinner with a fish hook deeply embedded in his mouth. I was so saddened I left the water for a time. Russel got in with the tow vehicle but the dolphins all left the bay. I went out in the zodiac and invited them to follow me back. They did. Again that forgiveness; I was humbled beyond belief.

18 September: Today on entry, I minded by own business and swam away from the dolphins but before long I was surrounded — they cut me off and redirected my course. Bentfin let me stroke his tail and we swam side by side. Poor Fishhook was eager to play. For an hour we swam together but those on the boat were anxious to leave for a scuba dive. With 40 around me I swam over. They circled me quickly, jumping out in a spectacular farewell.

19 September: Today we dived the wreck for the third time. As *Thor* was leaving port I noticed spinners all around the entrance. Some rode our bow a while. I had a strong feeling there would be no dolphins in the bay. After the wreck dive we went there. It was deserted. Heading home we passed the dolphins returning to their bay. Why did they choose to leave today?

22 September: The past two days were spent farewelling friends. Now I am alone. The dive boat dropped me at the mooring with the zodiac and I slipped quietly into the water. The dolphins were resting and I was careful not to disturb their peace. All around were larger groups than I have ever seen; almost double the usual number. Mothers and babies brushed past me as if I were invisible. I felt totally accepted and cried tears of happiness, finding it difficult to breathe through my snorkel at times.

After an hour of silent interlock the group came alive. Suddenly at my side was Rusty with his scratched up tail. When I dived he was right there, his belly against mine, but not touching. When we surfaced he would leap out and swim away, to reappear in a few seconds. He would swim upside down beneath me for extended periods. I was bumped and rubbed and squashed several times. The group increased and they chattered so loudly I could hardly believe that a few minutes earlier these dolphins had been so subdued. All around they were chasing and teasing one another. Hours passed and then I saw *Thor* waiting for me outside the bay. As I motored out to her the dolphins escorted me and then returned to their bay. This place should be called 'Interlock Island'.

23 September: Once again I was alone in the bay. Today visibility was poor but I could hear dolphins everywhere, high-pitched whistles and chatter. The first group to approach swam off into oblivion with me. The mothers allowed their babies around me today. They were so quick I couldn't keep up. Lumpy came in and singled me out to swim alone with him. I dived beneath him as he remained on the surface. I brushed past, feeling his warm underside. When I dived and spiralled up he circled very close, swimming overhead just before I surfaced. As I took a breath he waited patiently, one eye out of the water, quite still — and then we raced off together again.

Today there was an unusual degree of rough-horsing with dolphins deliberately ramming each other with their sides, open-mouthed nipping of fins and playful biting; much courtship and love-making, as usual, with large groups circling wildly, diving and cackling. I was quizzed with beeps and direct eye contact and made myself still in the water, just listening and watching. Time and again I was approached, 'spoken' to and beckoned to follow.

Regaining my energy I swam off with a large group. To my surprise, there was Rusty, my friend from yesterday, at my side. His manner was special and there was a lot of eye contact. He would swim off alone and return later. Time flew by and I became aware of the need to leave. For a moment I floated on my back, eyes closed, listening to the music of the bay. Meanwhile the dolphins formed a large ring around me, circling quietly in small groups. As I swam back to the boat the tempo resumed full scale.

24 September: Today four others accompanied me to the bay, but not with the same enthusiasm. When we arrived at 10.00 the dolphins swam over to greet us. I could hear their whistles above water. Such excitement. So many dolphins. I entered quietly as they waited for me. Thirty surrounding, we swam away from the boat. The first hour was difficult. My muscles were stiff from the day before and I got leg cramp. The group vanished into the blue as I massaged my leg. Then I became aware of a lone dolphin patiently swimming around me. To my delight it was Lumpy. He had a small cleanerfish on his belly. Occasionally he would rise half out of the water and slap down as if to rid himself of the little pest.

I swam with one group after another. There were so many new individuals; one with a fresh scar just in front of his blowhole — Newscar. It seemed that today the dolphins were putting on a show for me. They grouped and regrouped, tagging each other in wild pursuit. Occasionally, in their abandonment I myself was swatted and tagged. For a while my energy diminished. I snorkelled along enjoying the underwater terrain — large, meteor-like boulders; white patches of sand that bounce rays of light up to meet you, places of brightness contrasting with places of mystery. I found myself totally relaxed, my energy restored, ready for more. No sooner thought than done: my next interlock was even more wonderful than I ever thought possible. Different groups circled me: over, under, all around. As one set approached I swam to meet them. I blended with them and felt an incredible ease in my swimming. I was moving right along with them, but effortlessly, caught in their flow. Suddenly there was a surge and I was sandwiched between three dolphins. On my left was Newscar; on my right and beneath, two new individuals. The one on my right humped his back as I was pressed against him. I barely breathed and kicked only slightly, arms at my side. Eventually they broke formation to leave me rolling gently in their wake. By now those on the boat were getting anxious and I had to leave.

25 September: Today I went to the bay alone. There was no sunshine and I felt cold in the water and tired from the day before. My initial interlock was with a dozen dolphins. My energy output was weak and it seemed they knew it. They preferred to be among themselves. I did have an interesting beeping conversation with one dolphin. As we swam together he would beep at me a certain number of times and I would do the same back. When I initiated he imitated me. We 'talked' like this for ten minutes. I saw Lumpy briefly. He brushed past very close. My feet are so sore from my fin straps I have to wrap them with gauze and tape.

26 September: Today I was alone again, greeted by six dolphins. We swam for a while but then I had to stop. The thought of goodbye was in my mind and the lump in my throat restricted my breathing. I joined a pack and was questioned with beeps — the *same* individual as yesterday. When the rest of the group sped off in a burst of play, this dolphin stayed by me, joined by a companion with two remora on its belly.

A group of 20 dolphins sandwiched me in time and again, almost to the point of being squashed. A tail dislodged my mask; another tangled in my hair. All my aches and pains were gone, temporarily. My beeping friend returned several times for new exchanges. I saw dolphins open their mouths and regurgitate clouds of white particles into the blue water. As time ran out I swam reluctantly to the boat, to be joined by a crowd of more than 60: dolphins were everywhere I looked. I moved steadily but was cut off and impeded from rejoining the boat. We played a while and then I was allowed to go. My talking friend accompanied me to the boat and then departed.

27 September: Late this morning we left port in a different vessel. Just before arrival it broke down dangerously close to the rocks. I tossed out the anchor, jokingly calling the dolphins to come and help. Shortly after, I was called on deck — 60 spinners surrounded the boat. Swimming quietly, they stayed until a tow boat came to the rescue.

Today my interlocks were long and repeated. I swam among so many dolphins I felt I could never return to the rest of the world. I could not see beyond the group I was with — dolphins above, below and on both sides. They were touching me on either flank. I would lead the way and dive down. All followed and circled. I swam upside down. They covered me, blocking my way to the surface until I needed air, then a path was cleared. Rusty was with me today. I was so elated to see him. I burst to the surface and shouted for joy. The water was crystal clear and warm; the sky a brilliant blue. I could see forever around me, surreal scenes as large dolphins cruised quietly over the white sand, reflecting those piercing shafts of light upward to infinity. I could even see my shadow on the ocean floor as I swam slowly back to the boat.

Over the ensuing week Janet Nowak's sessions with the dolphins were disrupted by the arrival of a yacht squadron from Brazil. Although she found a few opportunities to be with the dolphins, it was not the same. People were chasing the spinners in dinghies and inflatables, and throwing food wastes into the bay. When she entered for a brief 'hello' there were no sounds. The bay was almost empty but for a few stragglers. 'How sad I felt that we had driven them from their playground. Two females with young came up to me. Together we left the bay. Mankind has done the dolphins an injustice. People don't understand the fragile balance. As yet there is no enforcement of the Reserve that Russel is seeking to establish.'

4 October: My last day on the island. I must farewell the dolphins. Chances are they won't be there because of all the confusion of the past week. The wind is rising and large swells are forming but I have to go, just one more time.

As we entered the bay I realised the dolphins had returned. My heart was bursting. I dived in before the boat was moored. Visibility was poor but I could hear all the familiar sounds. My special friend Lumpy was there to greet me and just the two of us had an exchange before I had to leave. I was escorted to the boat and said a prayer in the water on the way, wishing for their safety and that our understanding would grow. We have so far to go and yet the answers to our questions seem so easy to find. All we need is to open our hearts and learn the meaning of love. No matter how many miles separate us, when I close my eyes I am there.'

Postscript: With the efforts of Russel Coffin and his Marine Reserve Committee, along with the superb film Jack McKenney was contracted to make for them, and the resultant petitions of divers from all over the world, the Brazilian Government was convinced that more than half of Fernando de Noronha should be set aside as a marine reserve. For the time being, all entry to the Bay of Dolphins is prohibited until a sound management policy is developed which ensures that the dolphins' lives are not disrupted. There are four billion humans on this planet. Can we all claim the right to meet dolphins without causing them harm? Should we turn our backs entirely on their overtures of friendship? Or can we reach some compromise whereby through books, films and other media we can all meet dolphins, by proxy?

Credit: Jack McKenney

ee hundred kilometres off the Brazilian coast spinner dolphins, *Stenella longirostris*, provide a close parallel to the
amas encounters.

PLATE 30A

Credit: Jack McKenn

Spinner dolphins show similar responses to human encounter as increasing familiarity leads to play, one-to-one relationshi with humans, and demonstrations of love-making (opposite).

PLATE 30B

Credit: Russel C

PLATE 32A
In New Zealand a trio of common dolphins — *Rampal* (lower), baby (middle) and mother (top) — spend much of the lives in the Whitianga estuary, a rare habitat for this offshore species.

PLATE 32B
During playtime, the male *Rampal* invariably initiates gamesplay with the baby while mother remains aloof.

Credit: Steve

The Whitianga Dolphins:
Rampal, Mother and Baby

In 1980, a female common dolphin entered the tidal river at Whitianga, New Zealand, and it seems, in the mangrove upper reaches, she gave birth. Over the succeeding years this dolphin adapted to estuarine life in a rather unique way. Adjacent to the township, the river has carved out a deep gully which is twice the depth of the adjacent seabed. There, the dolphin found that each tidal change brought a profusion of prey fish within range of her sonar. Although river traffic was quite heavy at times, especially when the resort came alive in midsummer, the dolphin found she and her baby could avoid accidents and harassment by using a series of mooring buoys as her home territory. She would surface close to these without danger and if small craft bothered her, she moved from one buoy to another underwater, until the pursuers got the message and kept their distance.

Over the next five years, as far as we can reconstruct, four babies were born in the estuary and two survived. The original mother lost two; her first born, Nicky, grew to maturity, had a baby that perished, and left the estuary shortly after (spring 1985).

Just before Nicky departed another adult dolphin joined the estuary dwellers; subsequently we met this dolphin and called it Rampal because of its acoustic talents. From the outset, I followed the development of this situation with great interest, realising that it is unprecedented for this species, *anywhere in the world*. I maintained a file of anecdotes from people who interacted with the dolphins, but kept away for fear that publicity might overwhelm the dolphins before they were well established.

But in May 1986, Whitianga people urged Jan and me to come and document them for the public (TV and news media) because there were plans to build a large marina close to the dolphins' base which might endanger them. We drove down and held a public meeting, explaining other episodes of solitary dolphin behaviour and what we had learnt about free dolphins from New Zealand to Hawaii and the Bahamas. From discussion we were able to assemble a composite picture of the dolphins' life history in the estuary, as far as human observation could determine.

Our initial approach to the dolphins produced such an amazingly communicative response that it was front page news in New Zealand's major newspaper, the *Herald* and featured as major news items on TV.

We discovered that Rampal was extremely interested in exchanging sounds with us through our special underwater transmitting and receiving gear. Costing around $8000, this gear provides us with a two-way dolphin-human telephone. Rampal was prepared to spend up to four hours alongside our acoustic array, within inches of Jan's toes.

On the ensuing trip we were able to play the dolphin recordings of its own sounds which I had edited and manipulated with a digital synthesiser. Its interest in this situation was frenetic. But first, I should explain exactly how this unusual human/dolphin rapport began.

The afternoon of our arrival at Whitianga (May 1986), Jan and I met local conservationist Steve Hart on the wharf. Steve rowed us across the fast-flowing river mouth to *Betty*, a black steel fishing boat. At the bow three common dolphins were hovering just in front of the mooring buoy. We tied up to the buoy and hung back while Steve told us about the dolphins and we learnt to recognise each individual. As Steve pointed out where the marina was to be sited, the dolphins maintained their position in the swift current with little apparent effort, diving and surfacing continually. Sometimes they remained below for up to two minutes; sometimes just a minute — there was no pattern, but the baby always stayed with its mother and the other adult moved about independently.

For us, after years spent at sea, it was an incredible experience to sit there on the river and watch the open-ocean dolphins feeding in an estuary without a flicker of concern at our presence. A viewer on another planet seeing humans in orbit would not be more surprised. The mother had a rectangular nick in the trailing edge of her dorsal. Her baby had a perfect fin and the other adult (Rampal) had a nick in the leading edge of its dorsal, with a scar trailing back on one side. The fin was much more curved than the mother's.

Steve explained how the townspeople had become protective towards the dolphins and would not allow any harassment from visitors. A warning notice had been placed on the wharf. The dolphins could come and go as they pleased, sometimes joining up with the main school when it visited the adjacent bay, but always returning to the estuary. Steve told us stories of how the dolphins performed for children; how the old people would come and sit along the water's edge to watch the dolphins frolic at the day's end; and of the huge influx of tourists in the summer who enjoyed the ocean visitors. Just within the river mouth, Whitianga wharf is a busy place with a fleet of fishing craft plying to and fro and a regular ferry service to the old stone wharf on the opposite bank. The dolphins often accompany vessels in and out of port but they mostly ignore the boats when feeding. If the river ferry makes a detour to show visitors the dolphins, or if the tour boats come to see them, the dolphins seldom fail to perform swimming manoeuvres and acrobatics to delight the viewers. One day, ferryman Roger Eliot relates, one of the dolphins started to leap high alongside and splash his passengers. It went right around the ferry and never missed soaking one person. He himself was protected by the cabin windows but the dolphin positioned itself alongside the doorway and managed to douse Roger as well.

The dolphins would seldom interact with swimmers or divers. Typically, people swim straight towards them and, if given an opportunity, reach out to touch them. But one day Roger's 15-year-old daughter was swimming near the dolphins at slack water (the only possible time). She lay on the surface, her hands outstretched, eyes closed. She felt something touch her hand. A dolphin had rubbed its body against her fingers. She was thrilled and totally surprised as she had not been seeking contact.

When the fourth baby was born* the dolphins put on a great display of leaping and excitement down near the river entrance, just off Cemetery Point, then the mother brought her new offspring up to Ian, the other ferryman, and showed it to him.

---

* June or July 1985, just after Rampal joined mother and Nicky, her first born, who left shortly after.

As Steve rowed us ashore the dolphin trio quietly took up positions on our bow. Then Mike Elworthy went past in a power boat with a dog aboard. The dog barked at the dolphins on the bow. It always knew where the dolphins would surface in the opaque river. We had only to watch the dog's attitude and sure enough, the dolphins would ascend where he gazed. Mike steered by the dog's nose. We watched Steve row back across the river to his home, the dolphins frolicking in his wake.

We had been debating what the sex of the second adult might be. Later, Steve rang us and said it had taken up the male copulatory position with the mother, right behind his boat. This was the only clue we had to its gender but it could have been play-acting, as this dolphin appeared to have an auntie-type role towards the baby. For several months the question remained open.

The next day we set up our underwater communication gear on one of the vessels moored mid-river. Nearby, the dolphins were bobbing around the bow of *Betty*. By midday we were ready.

First I played some musical scales on the synthesiser that were amplified and directed out into the river through a powerful underwater speaker. In a moment the second adult dolphin was alongside the grey metal housing, gently nudging it with its beak, lifting it slightly on its head and rubbing it with its belly. A metre below the surface, the dolphin hovered beside the speaker for long periods, rising for a breath, curving in an arc and returning to the sound source. Through the adjacent hydrophone I could hear its high pitch signature whistles in a regular pattern and the staccato clicks of its sonar. At times I could see it release a large bubble-gulp from its blowhole — a familiar gesture to us, and usually made in a communicative context. To my delight I was now able to see it and listen at the same time; I could at last determine that this was *not* simply a by-product of sound production but more likely a visual signal. The hydrophone was *not* picking up any phonations when the dolphin released a gulp of bubbles, but when thin beads of bubbles emerged from its blowhole I did ·hear sounds and recorded them for later analysis.

Don Ross, the retired harbour master, had a yacht moored nearby and had developed a close friendship with the dolphins. He rowed over in a dinghy and showed us a bell and a metal bar he used to rap out a loud underwater signal. We watched as he demonstrated. Although the mother and baby showed no response (we learnt later this was typical) the other adult came to the stern of his dinghy, emitted a huge bubble-gulp and lay there, inches from the rim of the bell dangling through the surface. Then the dolphin arose, close to Don's hands and breathed, lying quietly on top of the water as Don spoke tenderly to it, asking the enigmatic question, 'What are you trying to tell us?'

For some considerable time the dolphin maintained a pattern of diving out of sight, circling around and rising just beneath the bell to release a bubble-gulp. I wondered if this could be a method of adjusting buoyancy as it faced into the current, no more than an arm's length below. But in due course all doubts as to the signal-intention were to vanish.

Then another local arrived. Ces Harrison also keeps a yacht close by. Ces told us that the dolphin always came to assist him when returning from sea by pushing the mooring float up as he reached down for it. On the other hand it would mischievously try to tip his dinghy over, nudge playfully at his oars and deliberately splash him with a side swipe of its tail.

When Ces shouted, the trio came over again, led by the second adult. It let out a trickle of bubbles from its blowhole and tried to lift the boat fender hanging from our vessel directly in front of Ces, who was cajoling it in a familiar, mock-bullying tone the dolphin obviously enjoyed.

The next day we set our gear up on *Baby Blue*, an old wooden fishing boat that offered better water access; a stern platform. As soon as I played some musical notes through the speaker, the trio approached and the second adult lay alongside it nudging it gently as he had the previous day. Overnight, I had been analysing my initial recordings and churning over in my mind the extraordinary responses of this dolphin, who was prepared to listen to our sound transmissions for long periods and to offer its own input at frequencies within our hearing range. But there was something missing. The muted voices of Don and Ces obviously carried through to the dolphin just below the surface — the hydrophone indicated this; but they were unable to hear the phonations the dolphin was making in response. Our hydrophone was picking them up and the recorder was amplifying them so that I could monitor the sounds with headphones as they were registered on tape. Now I had an inspiration. Among the jumble of fishing gear on board I found a length of plastic water pipe about a metre long. I suggested to Jan that she might sit on the stern platform, just awash, her legs dangling in the water. The tube would carry her voice down into the river and she could mimic the dolphin's bubble-gulp gesture prior to each sound transmission. Then I switched the recording machine to the public address mode and plugged in an extension speaker, placed on the stern near Jan's head. Now she could hear all the noises of the busy river and the dolphin's constant, high-pitch signature whistles and sonar click chains each time it approached. Jan said she would try and make the sort of noises a dolphin could readily mimic using its blowhole, but not having a voice box.

In her own words: 'Well, then it started to happen! If I blew a large bubble the dolphin would come up under me and release one too. Then I made a sound along with the bubble and the dolphin did *that too*. For the very first time in two days its bubble-gulps were accompanied by sound. I then made different noises and talked through the tube. I was so excited! The dolphin responded with amazing sounds such as we have never heard before in all our recording sessions at sea. What made it perfect was the feed-back system Wade had rigged so I could hear the dolphin's responses. We could actually have a two-way "conversation".'

The dolphin's behaviour indicated that it was as excited as Jan at getting a human response each time it completed a sound sequence. It wriggled and shook with seeming pleasure and made each ascent for breath briefer and closer to the boat; sometimes it would approach on the surface making raspberry sounds with its blowhole exposed, maintaining visual contact before diving alongside our acoustic array once more. We varied the pattern, either making the initial sound, once the dolphin was in position, or waiting for it to initiate and then responding ourselves. We strove to avoid the usual human-dominant situation because with a free dolphin, we feel it is essential to adopt a flexible, open-ended approach which is the essence of communication. In actual fact, it was more the other way around, with the dolphin able to manipulate us: rewarding responses it liked and ignoring those it didn't appreciate.

When Steve Hart entered the water to test the dolphin's reaction, he was inspected briefly and the acoustic game resumed. The session lasted for one-and-a-half hours before rain squalls forced us to seek shelter.

We resumed the next day. Jan simply shouted through the tube and the dolphin approached, greeting us with a bubble-burst, whistles and clicks. Now it would surface gently, right alongside the stern, look at us as it breathed, and descend. No longer did it explore the speaker, but just lay close to it as I transmitted through the synthesiser or cassette player. I had resources to mimic click chains, play snatches of classical flute music by Jean Pierre Rampal, or to create a variety of instrumental sounds at varying pitches, all of which, interspersed with Jan's output through the tube, maintained intense interest. Jan needed the instrumental support because using the tube was tiring. But the dolphin had the strong current to contend with and showed no weariness.

During the day a number of local people came out to see what was happening, including biologist Dr Stella Penny. She was delighted to sit on the platform holding a microphone, singing to the dolphin in the light rain that had us frantically covering our electronic gear. For quite some time as Stella sang to the dolphin, it stopped making its own sounds. There were no pauses in the singing to allow for its responses. But when Stella stopped and Jan resumed, the dolphin made an instant answer. It seemed to prefer a two-way exchange to a passive, listening role.

Several times it hit the hydrophone with its tail or mouthed it gently, perhaps wondering why it did not transmit sounds like the speaker. Occasionally it made a rapid sortie to snatch a fish from the current. We saw one leap from the water inches from its beak as it surged over the surface. But always it returned quickly to the stern of *Baby Blue* to renew the sound exchange. At times, correspondence between Jan's sound output and the dolphin's response was so closely matched we could only distinguish the two through close analysis using a special tape recorder that operates slowly.

As time went by the dolphin was out-doing Jan. It had mastered blowhole sound output (not a normal dolphin channel) to such an extent that Jan was easily surpassed. It showed how many nuances and varied patterns it could create, blending the air release noises with click chains and modifying the latter so that, at the bottom of the scale, they sounded like somebody running a fingernail along the teeth of a hair comb at varying speeds to produce changes in pitch. As the clicks accelerated they often merged in a rising tone, much like a phonation but with a distinctive quality.

Following another spell of flute music by Jean Pierre Rampal — Bach's Sonatas — we decided to name the dolphin Rampal as there was confusion among the local people between this dolphin and Nicky, its predecessor, who had a nick in the trailing edge of her dorsal. (There is a human tendency to name every fin-damaged dolphin *Nicky*.)

When heavy rain terminated this session, it had lasted two-and-a-half hours. We were exhausted. What about Rampal, down there, heading into the fierce tidal stream so effortlessly?

As we left the Whitianga Wharf in our VW van, Rampal was hovering around the jetty steps, an area he seldom visits. During our absence, Steve told us, the dolphin had waited by the stern of *Baby Blue*, spy-hopping and showing such excitement it even tried to over-turn a dinghy near the wharf.

Back in Auckland city we were contacted by television personnel eager to film this novel situation. We were glad to have an opportunity to document it on film, but would Rampal respond with a large group of people present?

The moment I played a tune on the synthesiser, Rampal returned to the stern of *Baby Blue* and stayed there for about two hours while the TV team filmed and recorded. At one stage, the dolphin came so close to Jan's feet its beak was only a hair's breadth from brushing her toes. Peter Thompson, the cameraman, put his rig in a housing. Holding it just below the surface as he sat on the platform beside Jan, he was able to get excellent underwater film of the dolphin exhaling through its blowhole in response to the voice tube. It came right under Jan's feet and released a very special bubble shaped like a doughnut — a perfect toroid, which trickled up around Jan's toes. A few scuba divers have perfected the art of blowing bubble rings, while filling in time during decompression stops — but it takes a lot of practice.

To round the session off, although the tide was racing by, I decided to get in the river with my ruby-red electric tow vehicle and see how Rampal responded. At full speed, dolphin-kicking to supplement its power, I could only just hold my own against the avalanche of water. So this is what Rampal has to contend with! The dolphin was close, slightly below me, on its side and in perfect eye contact. For a moment I flinched — afraid it might touch the whirring propeller blades — a groundless fear in a river full of outboard motors.

To our surprise mother and baby approached closer than at any time before. During our sound exchanges they would occasionally hover in the vicinity but it seemed as though mother did not want her baby to get too close. At times, we have seen that dolphin mothers find it hard to control their offspring once they get over-stimulated.

Seven months later, with summer at our disposal, Jan and I returned to Whitianga. We'd found the greatest difficulty in staying away, but did not wish to over-stimulate Rampal, perhaps in some way putting the dolphin at risk in the busy harbour. On the other hand, with common dolphins being hunted for captivity in our waters, it did seem important to investigate this alternative to dolphin jails as thoroughly as possible, and to see what could be learnt that might contribute to our understanding of dolphins globally.

This time our good friend, Ian Briggs (who had first reported the Monkey Mia dolphins), biologist and competent film-maker, was at hand with a TV team to record on professional video gear all that ensued. But again I had misgivings about expecting a wild dolphin to respond as it had done previously. I needed not have worried.

When the dolphin came over from its mid-river feeding zone it began exchanging sounds with us as if there had been no interval. The TV crew had three hours with Rampal alongside. This time we used Don Ross' yacht *Tere Manu* as base, with our inflatable lashed abeam as a diving platform. The film captures Rampal lying on the surface in an intimate tête à tête with Don, who is asking his usual question: 'What is it you are trying to tell us?' There is the dolphin looking at him and wheezing through its blowhole, the sun making diamonds on its lustrous skin.

When I hauled the speaker closer to the surface the dolphin rose with it. Inches from the lens, it nuzzled the speaker housing, prodding it gently with its beak as the camera rolled.

Shortly after this sequence, Rampal began to swim about excitedly. All of a sudden it swept past us, exhibiting a morsel of seaweed draped over its tail. Jan pointed it out. Her response seemed to encourage Rampal and several more passes followed, each time with the seaweed in a different place, finally on its pectoral, and side closest to Jan. Then it made weird little turns and twists close alongside in a high state of excitement as we clapped and cheered at its adroitness. We have seen this seaweed juggling act before, in two oceans and with four different species, and yet it is commonly said that dolphins have no power of manipulation.

By 5.30pm we were exhausted. The tide had slackened off and the dolphins resumed their station in front of the mid-river mooring buoys.

The next day the TV team wanted brief sequences from long range: on the cliff above the river and from the wharf opposite us. We had learnt by now that the best time for interaction was when the current flow was slowing down. Periods of full flow are peak feeding times, as all the fishes that enter the river system to feed out over the broad estuarine flats must pass the dolphins' sonar at the river's deepest, narrowest point.

A little reluctantly, we set all the gear up for the camera, just as on the previous day, conscious that this was an act. Rampal came over the moment we signalled readiness, performed blowhole noises on the surface close to Don Ross, passed beneath Jan, and released a gusher of bubbles. But he provided *no* underwater noises. Nor were they needed! The dolphin remained alongside long enough for the camera crew to get their shots and then took off, with no hesitation, giving a slap with its tail on the surface as if to say: 'That's it.'

We waved goodbye to the film crew — it was all over for them. The dolphins were feeding and we decided it was time for our lunch too.

At 4pm the tide slackened. Jan and I returned and set up the gear. Rampal came over at once, approaching us closely on the surface, its beak out, making long, drawn out blowhole sounds and looking up at us. I decided to try out a device called *Sea Voice* which enables divers to talk underwater: basically a rubber bladder attached to a special mouth piece. Rampal got very excited, swimming past making a sequence of three vocalisations in air, blowhole puckering and wheezing at the surface. It seemed to be trying so hard which did not seem fair — even the best trumpet player would find the human voice difficult to emulate. Unvoiced sounds seem more appropriate.

On the third day, Jan and I were alone with Rampal. I had been itching to try something which would take a lot of concentration and demanded a situation free from distraction. I wanted to play the dolphin a special recording I had made after analysing hours of tape recordings from the May trip. I had collected all the 'best takes' and put them through a special digital synthesiser, shifting them up and down the keyboard scale. Now I hoped to play this back through the underwater speaker, one phrase at a time, pausing to record any responses. To signal our readiness Jan tapped on an aluminium tube. The dolphin came over and immediately started vocalising beneath Jan and alongside the acoustic gear. Then it emitted a long, low reverberating sound such as we had never heard before, and rose gently to the surface where it made a sequence of blowhole sounds in air, as if demonstrating what had just been done underwater. At this point I took the opportunity of trying the playback tape in which each of the dolphin's most interesting phonations pass down the scale, slower and lower; then up the scale higher and faster, before returning to normal speed and pitch. For the next hour Jan and I entered an amazing space — the dolphin almost made the water boil, such was the intensity of its responses. At times it would surface beside us, raise its beak as if desperately trying to communicate as we in turn strove so desperately to understand.

At one stage I told Jan, 'Now I'd like to get a few photos like they got on camera with Don yesterday.' To our surprise, the dolphin seemed to grasp the situation because it lay there on the surface longer than at any time before while I took four shots in a row and then another sequence.

Eventually we were utterly exhausted. Concentrating on the complex task of manipulating the gear and staying tuned to Rampal's space had my head pounding. Then the dolphin did an incredible thing. It swam over to us on the surface. Slowly it curved its body like a banana, head and tail right out of the water, back all wrinkled. Even the lobes of the tail were curved upwards. It opened its blowhole wide, releasing air freely — and sank below. There it turned 180 degrees and sped across the river to its favourite mooring buoy. The clearest body language you could imagine to express, 'I am exhausted. That is it.' We were utterly amazed at the empathy it showed towards our own feelings. We felt satisfied yet humble. Out in the river Rampal lay resting on the surface alongside *Easy Rider* with mother and baby.

Before Jan and I returned to the north there was just one thing I wanted to try. As at the previous time, I wanted to terminate this special meeting with another in-water session using the tow vehicle. We crossed the river with it in our inflatable and Rampal came alongside. He then curved off towards the river mouth and disappeared. As I geared up alongside *Tere Manu* the dolphin repeated this manoeuvre. When I was ready to dive there were no dolphins visible in the estuary. Could those movements have been a signal to tell us they were going out to sea? It seemed a long shot, but we followed in the inflatable. Out in the bay the waves were rising with an onshore breeze and daylight was fading. We both felt we were pushing our chances a bit, when I saw a fin. Then a dolphin leapt. I rolled over the side. Jan passed me the heavy machine. As I slipped effortlessly through the water, the dolphin trio joined me: mother, baby and Rampal. For the next 20 minutes we romped and circled, spun and swooped. They leapt close alongside me on the surface. Below, they came in on their sides and looked at me — the closest I had ever been to mother and baby — an arm's length away. The water was murky from the out-going estuary, which made a close approach essential for visual contact. Once, while below, I very nearly collided with a rock and wished I had their sonar senses. Here I was performing as well as a human can manage, yet clumsy and almost blind, voiceless and almost deaf. How I wished I still had the use of our array of acoustic gear. Certainly, with expensive technology, I could miniaturise our gear so a diver could perform with it underwater, much as we had topside. In the United States much of the equipment can be purchased off the shelf, including a handheld sonar torch, but Jan and I have already become houseless with the cost of what we have already done.

It was surprising that as long as my body could endure the constant ducking and diving, strung out behind the machine like washing in the wind, the dolphins found me an interesting enough companion. Another time we would know what to do and meet them further out in the bay where the ocean is clear.

I thought as we travelled homewards that it was best to leave the dolphins to the midsummer influx of tourists and wait for a quieter time. Later, we could sail our catamaran down and this would provide us with better opportunities to continue our mutual exploration. Would we ever be able to answer Don's question? Why should a wild creature be prepared to spend hours at a time with no tangible reward, no hand out of food or security, just to exchange acoustic patterns with a species that has radically different sound apparatus? That question would lead us on. We had found a window into ocean mind that may provide insights into cetaceans in other parts of the world.

PLATE 33A
Just on dusk the Whitianga dolphins change their feeding behaviour as the fish schools disperse: from diving deep, they begin chasing fishes along the surface, leaping and splashing.

PLATE 33B
Pompal accepts diver Steve Hart in the water but avoids those who approach and seek contact.

PLATE 34A
Don Ross, ex-harbour master, has a special friendship with *Rampal*.

PLATE 34B
Surfacing to breathe and then submerging, *Rampal* engages us in prolonged sound exchange sessions, recorded with vic
and acoustic tape.

PLATE 35A

Rampal creates a massive bubble burst to signal his readiness to communicate. Jan responds with a tube, then a microphone, repeating and varying the syllables uttered by the dolphin.

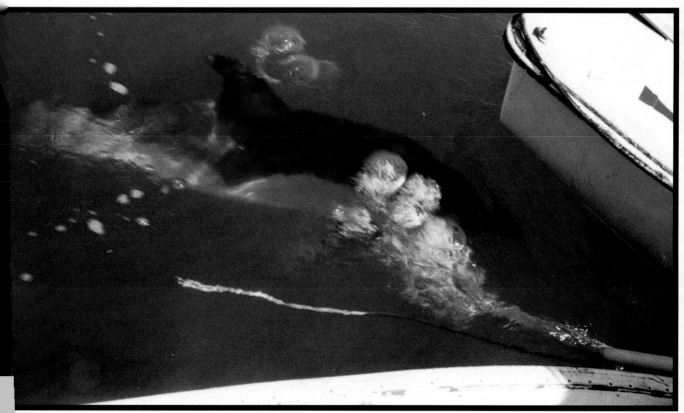

PLATE 35B

Credit: Steve Hart

Mutual 'bubble gulps' function like the ringing of a telephone. A metre below the surface, our hydrophone and speaker receive and transmit novel sound patterns between dolphin and human.

PLATE 36A

Credit: Martin Jack

In the busy Port River, Adelaide, in 1988 the bottlenose dolphins interact with humans in several unusual ways. *Billy* joined by a dog Rastus, accompanies horses exercised behind trainer Sandy Sandford's boat — a four species encounter.

PLATE 36B

Credit: Martin Ja

Rampal is so different from the solitary dolphins on our files, but that may be a consequence of the situation. A fast-flowing estuary is not conducive to physical exchanges and this led us to explore the more abstract, acoustic capacity. Our rapport is rather like that of a human mother and baby as they exchange sounds and attempt to match vocal apparatus, in the 'da da da' games — the beginnings of speech. But how far can this go between species with little in common other than the habit of communicating?

In Hawaii, Dr Louis Herman has made a long-term study of the ability of captive dolphins to understand commands given in both acoustic and sign languages. His paper 'Comprehension of Sentences by Bottlenose Dolphins' concludes that they do have the ability 'to process both semantic and syntactic features of sentences'. In short, they obeyed a variety of commands given in two different modes with an astonishing degree of flexibility. Their performances are compared with tests given to young children and found to be 'well developed'. The experiment avoided the area of language production and the dolphins were motivated by fish rewards which were part of their daily food intake. Dr Herman concludes with the observation that further research is needed to explore the boundaries of dolphin competency, including their ability to produce language.

Rampal offers some indication that there may be avenues to explore with non-captive dolphins. Motivation by rewards less basic than food may lead to situations more conducive to a two-way language exchange. Such work might take many years and great expense. How frustrating that Jan and I had to wait.

When the peak of summer passed and the river traffic resumed its normal business after the chaos of holiday time, Jan and I prepared for our third expedition. I felt I could not complete this book without further investigation even though I realised that what we had to learn from Rampal might take many more years. I needed to get better quality recordings for laboratory analysis, without the distortion of river noise caused by current flow and marine engines. Perhaps, with our own vessel as base, I could record at slack water and at night. Besides, we needed to know much more about the dolphins' basic living patterns. To make progress, we would have to live on the river for an extended period and observe their activities continuously. And we just had to discover the gender of Rampal.

In late February 1987, we set sail on the longest voyage we had ever made with the catamaran — 113 sea miles southwards. Jan and I have few illusions about our sailing ability. We are divers first for whom boats are a means of getting to a goal. Only dolphins could induce us to venture so far with so little seamanship.*

Before leaving, we had the help of our dear friends in giving the catamaran its second major overhaul in a decade. We bought flares, a CB radio and a new motor for the inflatable. A major diving-gear firm presented us with a full range of the latest scuba equipment, no strings attached.

After two days of strenuous sailing we were once more beneath the towering white cliffs within the river entrance gazing at the strange wind and water sculpted rocks with their frieze of pohutukawa trees, some whitened with shag (cormorant) droppings from the colony nesting in the upper branches. As we secured to the mid-river mooring, mother dolphin came past our bows

---

* We had three near-disasters on the voyage home.

and blew a raspberry through her blowhole. A greeting or a warning? Her offspring poked its head out and looked at us. And so began an 11-day period in which we lived close to the dolphin trio, a unique opportunity to observe their daily lives and they, ours.

Our first surprise was to discover the dolphins were not remaining in the estuary at night, as was commonly assumed. During the day, as the tides raced in and out (we were there for the period of king tides), the dolphins remained close together, diving in unison in front of permanent moorings in the river throat, feeding constantly for up to four-and-three-quarter hours on end. In one day we saw them feed for seven hours in two equal sessions with just an hour's relaxation at the top of the tide. Dives would last up to two-and-three-quarter minutes but were usually about half this duration. Recovery periods intervened in which they made a series (3-5) of brief, shallow descents. Rampal appeared to have slightly less capacity and sometimes the baby showed a little more than its mother. We noticed an interesting pattern: the greater the current flow, the deeper the dolphins went during the recovery period. In near slack water they simply remained on the surface for around 30 seconds between feeding dives.

A friend offered a good explanation with reference to the science of fluid dynamics: when a fast-moving object breaks the water surface, drag increases. Minimal drag occurs when a spindle-shaped object is *three times* its own diameter below the surface. Just beneath, drag is at its worst, decreasing as the object breaks through.*

Clearly the Whitianga dolphins have mastered the principles we learnt from wind tunnels and hydraulic tanks, and some of their behaviours can best be understood with reference to this science.

When dusk coincided with slack water their co-ordinated hunting pattern changed. As the sun slipped behind the Coromandel Range the mother thwacked her tail four times on the water and the dolphins began to dart about just beneath the surface, bursting out in rapid lunges as schoolfish skittered over the surface in front of their jaws.

It seems probable that this change in the dolphins' hunting method was dictated by the night-time behaviour of their prey. With the diminishing light, fish schools break up and disperse near the surface. For dolphins using sonar in the turbid estuary, day or night would make little difference, but fishes are visual creatures and suffer much less predation after dark — except for the sonic 'torchlight' of the dolphin. For several hours after dark, the dolphins pursued individual fish all over the estuary, gradually shifting their operations out into the adjacent bay. By around 10pm the river was silent. We found they did not return to the river until morning, and even then, not until the tide was really on the move.

One morning while the dolphins were still out at sea during slack water, I donned scuba gear and descended the stout mooring cable to the riverbed ten metres below. To my surprise, it was a jumble of car-sized rocks densely encrusted with mussels. Through the murk I saw several species of reef fish darting to and fro among the rocks, and silvery sprats hovered close to the cable for shelter. My friend John Calder, diver and fisherman, told me he once saw a school of about two tonnes of jack mackerel hovering in mid-water in this area — dolphin prey. I tried to envisage the dolphin trio down here when the tide was sluicing by, their echo pulses bouncing off school

* Heinrich Hertel, *Structure, Form and Movement*, Reinhold NY, 1966.

202

fish whisking past among the shadows of the rocks; two minutes below, a lungful of air, several recovery dives and below again, plucking prey from the water wind hour upon hour.

Our next surprise was to discover the dolphins had resting areas further upstream from the main feeding zone and that the trio separated during these times. Rampal would cruise slowly around one particular mooring buoy, for the most part on the surface, occasionally descending for a few seconds. A hundred metres away mother and baby had their resting zone between two moored fishing boats, but in their case they stayed below most of the time, with descents averaging around ten seconds, rising for two seconds to breathe. It is believed dolphins 'sleep' with one half of the brain at a time, remaining alert visually but shutting down their acoustic capacity. The longer we studied these resting patterns the less regularity we found. Some days they would all rest for up to two hours, others for as little as 15 minutes. They might have two rest periods and occasionally Rampal would rest longer than mother and baby.

As the days went by, watching from the cliff tops, we were able to observe the dolphins feeding out in the adjacent bay. We came to understand that these dolphins were not resident in the estuary as was commonly believed, but included it in a large feeding territory of which the farthest limits we have not established.

While feeding was most intensive during peak current flow, play sessions tended to be at dusk or low tide, slack water, but like all their activities, nothing was consistent or regular. Socialising would often last for up to two hours. Invariably Rampal was the initiator. He would approach the baby in a frolicsome manner, perhaps nudging its tail up just as it was about to breathe, so that its head went under; or inducing it to play the dolphin-roll-over game in which one rolled upside down over the other's back. There would be synchronised leapings, cross beak spy hops, chasings and mutual beak proddings while mother hovered nearby, 'not amused' it seemed. Only once did we see her join in a play session.

For us the greatest surprise of all was the way in which we learnt Rampal's sex. By the third day of our expedition we were burning to discover this.

A scientist who published a report on the dolphins for the marina developers after a brief visit, stated that the second adult was 'almost certainly a female'. This was totally at odds with our own intuition, based on a pattern of small clues.

On the third afternoon, up on our bows I had just explained to the Calder family how to sex a dolphin if they could glimpse its underside. I had drawn the patterns of genital openings for them, when suddenly, as a launch roared past, Rampal streaked after it and began to treat the stem as if it were a female: upside down, body curved around it, pectorals embracing the sides. We all stood there, rigid with astonishment at this unusual display while Rampal repeated the performance five times. But still the penny did not drop.

Next morning as I sat in the cockpit sipping my coffee, Rampal came in from the sea straight up to me, made eye contact and dived beneath the mother who was on the surface by the hull. Just beneath her body but slightly to one side Rampal lay upside down and extended his penis twice, very deliberately, from the slit in his belly. (The dolphin penis has muscular control.)

This did not appear to be a sexual act and the mother showed not the least response. We had seen no courtship behaviour in any of our past or subsequent observations. From then on, we could have no doubt as to his gender but the whole episode left us more than a little puzzled as to why it should have occurred as it did.

Later that morning Rampal approached us for the first of a series of communicative exchanges such as on our previous trips. But this time so many things were different. No longer did we signal to Rampal with the ringing of a bell or some acoustic output. From the outset he made it clear he would come of his own accord when he was ready and from then on, no signal from us would entice him over. He chose the time for the interlocks excellently — at periods when the current flow had slackened — and he took the initiative in all the exchanges. No longer did he explore the underwater speaker or bother with irrelevant games. He seemed to have grasped where our capacities lay. No longer did we have to cope with an acoustic array vibrating noisily in the current, nor did he extend the sessions to the point where we were exhausted. Our richest exchanges were just under an hour in duration, sometimes twice in a day, and always in perfect conditions.

Because the river was now discoloured from rain sweeping down from the foothills, we were ignorant of his presence beneath our stern platform until he made it known. Our acoustic array, broadcasting per the p.a. system, made us aware of his clicks and whistles in the vicinity but his actual position was a mystery until a huge gusher erupted beneath us. Jan noticed that the initial bubble-burst was always the greatest. 'Like an atomic bomb cloud,' she remarked.

Rampal then established a regular pattern of exchanges which made it easier for us to respond appropriately and to analyse our tape recordings later. Each bubble-burst would be acknowledged by Jan with a matching unvoiced burst from her tube. Then, as regular as clockwork, Rampal would make a new sequence of blowhole sounds. As time went by these became increasingly complex and prolonged. To let him know that we had heard him, I would make an attempt to copy his sounds or make some variations of them, weaving together previous patterns he had taught us into a new synthesis. This appeared to stimulate the dolphin to more intensive efforts and the session became rather like a two-way telephone conversation except that neither party could communicate much more than joy and creative spontaneity.

I wondered whether a language bridge might eventually be created and made a few attempts in that direction, but became aware of some technical barriers. Without seeing his sounds on the screen of an expensive spectrographic analyser, I could not fully appreciate the complexities of the patterns Rampal made. We were visual creatures encountering the world of those that inhabit an acoustic reality. Moreover, on Rampal's side, there could be another problem. Scientists believe that dolphin hearing is impaired at low frequencies. We may need to explore the possibilities of using frequency shifting apparatus so the dolphin can hear our sounds within his usual hearing range. There are many problems but we may have taken a few steps across the bridge toward another intelligence; a mind alien to our own, and one where our expectations act as blocks to understanding; but a mind which has already demonstrated such benignity, and capacities exceeding those of any creature we humans have encountered in Universe. While it is essential not to vaunt them to the heavens or to regard dolphins as 'humans with flippers', one thing is certain: in the past and as this is being written, we have treated cetaceans in ways which do no credit to *Homo sapiens*, and which impede our progress towards a deeper understanding. Just as Africans were once exhibited at fair grounds in England, next to a booth where a man ate a live tom cat, dolphins and whales are plucked from their social groups to spend their lives in concrete cells for our entertainment or research.

While the analogy with African slaves is not accurate, I do think the mistake we have been making is of a similar order.

Time for another leap forward, says Rampal!

PART FOUR

A Window on
Whale Mind

*Moving through a dim, dark, cool, watery world of its own, the whale is timeless;
an ancient part of our common heritage and yet remote; aweful, prowling the
ocean floor . . . under guidance of power and senses we are only beginning to
grasp.*
Victor Scheffer

Global Survey of Human/Whale Encounters

Can humans get close to whales? Can we gain insights into the cognitive capacities of the larger cetaceans from interacting with them in the wild?

Sometimes we have no choice. The story of BW, a solitary beluga, parallels our file of lone dolphin encounters. We received this narrative from Kevin Vanacore of New York:

On the morning of 21 May 1985, Kevin and his wife Claudia, were cruising inside the breakwater of New Haven Harbour in their inflatable. Fifty metres off the bow they saw the finless back and pale grey form of a beluga whale *Delphinapterus leucas*. Kevin put his outboard in neutral and the whale surfaced at the stern. Vertical in the water, its head within inches of the underwater exhaust, it appeared interested in the bubbles and sound from the engine. As they drifted with the motor in neutral the beluga would ascend and descend while remaining upright. Often it released large bursts from its blowhole as it peered into the engine's lower unit. When Kevin stopped the motor the whale descended and then reappeared alongside the boat. They saw that it was about three metres long — a juvenile white whale.

Four times the beluga approached the silent motor to release bursts of air, as if mimicking the engine or prompting it to produce more bubbles and sound. Frustrated, the whale came alongside the zodiac; upright a few metres from Kevin and Claudia, it repeated the expulsion of air from its blowhole, just below the surface. Getting no response it came up and shook its head violently to and fro so the melon seemed out of control, and it persisted in this with loud squeals from its blowhole.

'This led us to believe it would be more at ease with the motor running. We did this and the whale returned to its previous position, inspecting the bubbles at the prop. As the beluga became more accustomed to us it began passing under the zodiac, its echolocation clicks clearly audible through the resonant hull. It began gently brushing the hull as it passed below, as if enjoying the feel of the fabric. This contact continued until it remained stationary, rubbing against one of the zodiac pontoons.

'Then it allowed our hands to pass along its side as it swam slowly by. Its swimming slowed to the point where it was purposely positioned for contact with our hands. Then it remained still, allowing all areas of its body to be stroked — with the exception of the flukes and around the eye. To our surprise it was receptive to being rubbed on the melon and even near the blowhole.

'Forty-five minutes after the start of the encounter the beluga began nudging a small rubber boat fender tied to the side of the zodiac (an inflated 12 inch long tube). I untied it and let it drop to the water, attached by a metre of nylon cord. A metre below the surface and about two metres from the fender, the whale in a vertical position, swung its head from side to side emitting an explosive sequence of echolocation clicks. With its lower jaw tucked down towards its chest the beluga scanned the object for almost a full minute. It then swam beneath the fender, prodding it a couple of times with its melon.

'Still vocalising profusely it rolled on its side, grasped the nylon line between its lips and swam off with the fender dangling along its side. After a short display on the surface (Kevin included four photos of this episode), it brought the fender back alongside and released it as it dived. I retrieved the fender and tossed it ten metres from the boat. Repeating the rapid vocalisation and scanning the fender, the beluga again picked up the line and swam to the boat, releasing it within Claudia's reach. Then it waited for our next move. When Claudia did not toss the fender away again, the beluga repeated the behaviour of violently shaking its head and melon, vocalising wildly with its blowhole awash. The squealing ceased when the fender was put back in the water.

'Then a new twist was added to the interaction. The beluga again retrieved the fender but this time it submerged with it. When it reached its desired depth it released the fender, along with a large amount of air from its blowhole. When the fender broke the surface the whale immediately grabbed it and repeated the behaviour a *dozen times*, towing it around and skilfully avoiding our reach.

'Then it added one final variation to the game. We had expected the float to pop to the surface with the whale close behind when it dived again with its prize. Instead we felt a thud on the bottom of our zodiac. I assumed the beluga had simply brushed the underside, as previously. But it surfaced *without* the fender. To my astonishment, the whale lay beside the boat watching us intently. For several minutes it allowed us to stroke it. Again it released a great burst of air from its blowhole. I wondered if it were trying to coax me into the water? Then it sank slowly and resurfaced with the fender. It had been concealed beneath our hull! The whale allowed us to remove it gently from its mouth.

'After two hours of intensive interaction, the whale's attention was caught by a lobsterman nearby. I motored back to shore, with a feeling that we had been privileged in experiencing these rare moments. We had offered no food rewards. Perhaps tactile reinforcement contributed to the length of the interaction. (Kevin is a horse trainer.)

'Normally a gregarious animal, the lone beluga appeared to use physical contact with us and our boat to supplement its solitary existence. The duration of the exchange also led to a chain of mimicry and seemingly inventive behaviours. The beluga's response to our lack of response was amazing. When we couldn't fulfil its need for interaction it appeared to taunt us into trying something new.'

Beluga whales, an Arctic and sub-Arctic species, are not commonly found in Long Island Sound near New York city, but in February 1985 fishermen and tug skippers had become aware of a solitary juvenile they called BW. By April reports were coming in of it hanging about fishing boats, possibly feeding around their gear. BW visited popular beaches and alarmed two scuba divers who met it at night. It was extensively photographed and filmed alongside a yacht and it frightened the daylights out of windsurfer, Charles Goetsch: 'My breath froze. Ten metres away a silver shape heading straight for me, just under the surface . . .'

On the verge of ramming him, it lunged out and dived beneath the board to surface behind him, snorting. Close alongside it sprayed Charles with exhalation and sank beneath the board. For the next half hour it frolicked around the windsurfer making close eye contact, breaching and diving.

'It would submerge a few inches and drift under the red plastic patch of my sail where it would roll over and just lie there as if contemplating the sky through the strange lens.'

Once it surfaced behind Charles and started pushing the sailboard with its head. When Charles tried to pet it, the whale appeared receptive but then, with a final leap, it left.

Over the summer, other reports tell of the whale pushing an inflatable, with parents and children aboard, for an hour and accepting gifts of fish.

Alas, on 13 May 1985, five kilometres south of New Haven, floating belly up, a dead juvenile female, presumably BW was found with three bullet wounds in its body. There was a great protest from Connecticut residents and the story was carried worldwide that the State Governor offered a major reward for anyone able to find the perpetrators.

A parallel situation had occurred in the same area in 1980 when Bella, a solitary beluga, frequented the bays of Long Island to the delight of local people. She disappeared soon after attempts were made to capture the friendly whale by people from an aquarium.

In September 1982, Elsa, a young female orca, spent nearly a month in Provincetown Harbour, Cape Cod. Most days she played tag with a woman in a kayak. She bumped windsurfers off their boards and approached people almost within arm's reach. When she had eaten enough of the fish thrown to her, she would parade slowly at the surface with one hanging from her mouth. One day she followed a fishing boat to sea, the same one she had arrived with.

Physty: The Sperm Whale
that Visited New York

Close to important whale feeding grounds in the north-east Atlantic, the coastline between New York and Boston features strongly in our records of human/whale encounters. Public awareness of cetaceans in this major population centre rose even higher in 1981 when an eight-metre male sperm whale, *Physeter macrocephalus*, stranded on Coney Island, New York. Towed to an empty boat basin at nearby Fire Island, the sick young whale was cared for by a resourceful team of experts and well-wishers. During a herculean ten-day struggle to find and administer the appropriate antibiotic for its congested lung, the whale almost died twice.

Among the helpers was a United Airlines pilot, Bill Rossiter, whom I later came to know well. Closeness to a whale was not new to Bill. He had already experienced the friendly overtures of grey whales on the Mexican-Pacific coast. But five consecutive nights, caring for Physty, as New Yorkers fondly dubbed their sperm whale, offered Bill unusual insights into its capacities.

'There was no question that Physty related to people. He seemed to recognise certain individuals and voices. Occasionally he would flex up, raising the "acoustic window" area of his head above the surface and clicking directly at people. At night he often remained close to the few present, clicking more frequently and appearing to listen intently to voices speaking or singing to him.'

Bill was amazed at the whale's tolerance of painful medical attentions: scalpel blades, syringes and blowhole swabs. 'As a vet sought a blood sample for analysis, the knife cut deeply enough to cause the whole body to quiver, but the whale endured the pain without lashing out.'

Following his first night's vigil Bill wrote: 'In the predawn chill of Easter Monday I ran around the edge of the basin, all alone with the whale. His clicks followed me, pulsing into the dock so strongly I even heard them echo and felt them through my shoe soles. During the night he often approached and allowed people to stroke him. He came back for more. To feel the intense pulse of sound strike your palm, inches in front of his head, was strong evidence for the theory that sperm whales may stun their prey with sound.'

The eighth day began with sadness. Three penicillin injections given on preceding days seemed ineffective. Physty was close to death. 'For long hours life was present only through eye movement and weak breathing.'

By now the daily crowds of awe-stricken New Yorkers were several thousand strong. The previous Easter Sunday a hopeful throng of 8000 had watched attempts to feed the whale squid — but his jaws had remained closed. A great pall descended over the watchers. Many of the scientists had left to prepare for a necropsy and disposal of the 12-ton carcase.

As night closed in the crowd remained. Physty was not going to die alone. 'We began a vigil shared by this huge hushed crowd. Attendants stroked and consoled Physty. Laurie O'Gara kissed the great head and felt a deep

212

response. Tears were shed as the whale grounded with the tide.' Desperate measures were needed. Attempts to administer a new antibiotic with an improvised three-metre stomach tube had failed. Then Michael Sandhofer, a commercial diver and member of the group 'The Seas Must Live', came up with a daring plan. Manning the safety boat, Bill Rossiter watched at close range.

'Who would have guessed a man would swim up to 25,000-pound sperm whale in the dark and after strong and gentle touches, hugging and talking, insert a squid crammed with chloromycetin tablets into the whale's mouth — in the dark! He was visible to the crowds on the dock only by his white woollen cap. For an instant Mike's hand was caught as Physty's jaws closed on the squid — a mistake, no harm meant.'

The next day in heavy fog and rain, Mike made a further feeding of medicated squid. For the first time Physty showed a friskiness, swimming with powerful strokes, Mike riding on top to avoid being swept away.

That night, his fifth bone-chilling, sleepless vigil, Bill Rossiter became conscious that the whale was ready to go.

'When I talked to Physty he came closer and clicked as before. Finally I asked, "How do you feel? What do you want?". The massive body turned almost immediately, and with three strong pulses of its tail fluke Physty accelerated, turned sharply and coasted directly at the net we'd strung across the entrance to the boat basin. For the rest of the night he never repeated this movement.'

On the tenth day, with several thousand New Yorkers cheering 'Go Physty, go', the whale was escorted towards the sea by 11 boats. It swam strongly at around five knots and began diving as soon as open water was reached. The whale entered the ocean quickly and was last seen by the Coast Guard flotilla about eight kilometres offshore where it dived and was lost to sight.

For Bill, and so many involved with Physty, 'the whale gave us more than he took'. Some 50,000 people had crowded the dockside during the ten-day epic — children with *Get Well Physty* placards; people with flowers; old ladies with flasks of chicken soup and bottles of holy water. Many stayed all day; few left unaffected. Nightly it featured nationwide on television news and each day in the print media. Eventually television coverage extended as far as New Zealand and the Peoples' Republic of China. Wrote *Time* magazine: 'Each year thousands of whales run aground and die along the beaches of the world . . . this was the first time that a stranded sperm whale had ever been saved.'

Through caring for a sick whale, the whole world felt more worthy.

Close Approaches

Many people find it hard to accept that the mightiest creatures on this planet can be as gentle and curious towards our species as any dolphin. But considering that a dolphin is really a small-toothed whale, it is not surprising that close approaches have now been recorded, right around the world, among the more common of the 76 species of toothed and baleen whales.

With the decline in whaling, some cetacean populations are showing signs of resurgence in their former feeding and nursery areas. Wherever such aggregations of whales are accessible to man, the possibility for close encounter is much greater. As mutual fear diminishes, people and whales are coming within visual, auditory and tactile range — close enough to see each other, hear and even touch. In such situations, individuals of each species have signalled a readiness to accommodate the other. Practical moves have been made; for the first time in history many whale grounds have been designated sanctuary status. We now have the technology to meet whales on their own terms, listen to their voices and transmit their images around the planet. Gradually we are being taught, by reward and deprivation, the appropriate ethics for meeting leviathan in a benign setting, and we are overcoming vast differences in body size. And so, over the past decade, Project Interlock files have accumulated accounts and photos of close approaches involving 14 different species of whale from beluga and orca to the mighty finback and sperm.

In establishing interspecies trust, we must not forget that man comes with one hand behind his back. On a personal level we may accommodate each other but as a species we are competing for ocean resources: squid, krill, school fish and unpolluted territory. In monopolising these resources we hold the future of the whales in our hands. Their only hope is that an increasing appreciation of their capacities may motivate humanity to give them space — and so enhance our own survival.

Many scientists flinch at the term 'friendly whales', yet cetacean curiosity and consistent non-aggressiveness towards humans is increasingly obvious. Others, more venturesome, would suggest that we may be discovering the term 'friendly whale' is not an anthropomorhism, and that between individual humans and cetaceans the possibility exists for a relationship based on trust, mutual respect and curiosity, leavened with a shared appreciation of playfulness. Is it too much to extend the concept of 'friendliness' to members of another species; even our closest brain rivals on the planet? All I can hope is that the experiences of those at the interface will be considered; people whose researches or leisure pursuits have brought them close to leviathan.

Our global survey of close approaches follows an organic, associative pattern, expanding encounters in the Boston-New York 'hot spot', and radiating from there to other American coasts and beyond — the Indo-Pacific, the Mediterranean and southern England.

Stub — In 1983, for the second time in my life, I journeyed to the north-

214

eastern coast of the United States on energy generated by the local enthusiasm for whales.

At the Whales Alive Symposium in Boston I was delighted to meet a man who seemed like a Northern Hemisphere counterpart. For about as long as Jan and I have sought to understand the friendly overtures of dolphins, Bill and Mia Rossiter have done likewise with whales. Of course, in both cases we would be just as happy if the Doaks were regularly meeting whales or the Rossiters, dolphins. But biogeography has determined that the nor'west Atlantic, where this airline pilot spends his leisure hours, is a summer feeding ground for humpbacks, minke, fin and right whales.

It is a nice irony that along the coast from where the Yankee Whalers hailed, whales are re-establishing their populations, and this time in amazing harmony with man. Whale watching is a non-consumptive, rapid-growth industry along the US Atlantic seaboard (and the Californian coast).

During the conference, 200 delegates from 28 countries embarked on a whale-watching jaunt. When two humpbacks, Stub and Binoc, surfaced right alongside our ship and made eye contact, and Pegasus gave a joyful display of leaping with Sir Peter Scott and Walter Cronkite among the gape-jawed, lens-clicking throng, I was convinced that mutual trust, curiosity and play were being demonstrated on the Cape Cod whale grounds.

My conviction deepened when Bill Rossiter wrote to me later. Five days after our encounter with Stub, Bill took his inflatable out in the same area on Stellwagen Bank:

'I decompressed from the conference by sitting alone 28 kilometres offshore, drifting in a flat calm — two of the nicest days of the year, listening to whale sounds: blows, snores and tail slaps. I read your paper ("Playing with Whales") and others, in what had to be the most appropriate of settings.

'Then Stub, a large adult humpback, *Megaptera novaeangliae*, and Sirius, a yearling male, began to manoeuvre close to my boat. For the next 48 minutes Stub and I focussed on each other while Sirius behaved either in boisterous play or in a somewhat bored fashion, trying to seek attention or to induce Stub to move on.

'Head to bow, parallel to the boat, Stub kept his eye a metre from my face. Eye contact was prolonged and repeated as Stub manoeuvred in a tight area, his head occasionally grazing the boat but always slowly and carefully. With his eye as close as 20 centimetres away, I tried opening and moving my fingers and fist. Stub's eye movement and fixation showed his interest. Then I began to mimic his pectoral fin movements with my arm. Once, to put his right eye where his left eye had been, he sank his tail and arched his head up in a steady flowing series of moves that ended with his right eye a metre away. The control was incredible as his eye and rostrum passed close to my head, as if there were a wall there.

'When his upper jaw was against my bow, I put two finger tips on his head. In response he began to duck and move forward; then stopped and reversed. I made no attempts to touch him out of his line of sight. Several times his pectoral fin would rise slowly to within centimetres of my hand. He seemed timid and hesitant to touch me. Always he moved with incredible caution, attaining positions relative to me and controlling them to the centimetre. Within a sphere the diameter of his body length, he sank and rose; rolled with pectorals splayed or tucked in.

'I say "he" because during belly-up passes I noticed no female characteristics, although Stub and Sirius behave like cow and calf.

'Adrift in my boat, I had not invaded the whales' space intentionally or tried to restrict their actions. At no time did I expect too much of Stub in maintaining control. There could be mistakes, but I felt no concern. Yet, at some time, every part of Stub was within a metre of me — except his tail fluke. On one slow pass his fluke hovered two metres below, held down at a 30-degree angle. Often his head and back were only centimetres away.

'My eyes brought rich memories of worn and scarred skin; shapes and textures of mouth, throat and belly; tubercle hairs, barnacles and cyamid parasites; the great quivering expanse of back; but most of all, Stub's eye, its brown-grey iris, dull cornea and surprising degree of rotation and projection. Behind that eye there was a mind. He seemed to be sensing the boat and me with steady, prolonged approaches, orienting various parts of his head at me. What was he sensing with? I could always see at least one eye (he favoured the left), suggesting a wide visual field. The posture of his head suggested listening.

'Wearing mask and snorkel I put my head over the side. Directly below me Stub made several passes emitting strings of bubbles from his blowhole. A humpback seems to pass endlessly from that range. Later I noted three massive exhalations underwater. His breathing was unlike the normal swimming pattern. Most blows were made by simply flexing up.

'Perhaps I should not have left my outboard running. When I made sounds through my snorkel there was little response but he showed curiosity when I tapped the hull. I was acutely aware that I was missing a great deal because of my limited sensory capacity.

'All the while Sirius was active in the vicinity. He stood on his head, slapped his fins, spy-hopped, tumbled and was often obscured in surface froth. Several times he tried to push Stub away. He behaved like a bored puppy and even slapped the stern pontoon twice with his tail, more out of demonstrativeness than with any real force.

'After 25 minutes with my head in the water I stood up for a better perspective and to try other ideas. The encounter lasted another 27 minutes and ended two minutes after I yielded to a rational prompting, and brought out my camera. Unquestionably, that broke the contact. A change in mood? Stub came up to the stern with Sirius nearby and then they surged away down, with the strongest movement of the whole encounter. Two minutes later they surfaced 200 metres away and swam off. I did not follow.

'Why didn't I get in the water? I've tried it a few times and never found it productive. I'm out of my depth in that I'm rarely able to see the whale (the whale ground is green with plankton), unable to hear well and distracted by cold. Leaving the boat so far from shore could be unwise. If I use a tow line I may get entangled with the whale. Still, with individuals I've met before, I plan to try it.'

Sockeye — Two years later Bill sent us an underwater photo of a humpback eye and the base of its massive pectoral fin. With it he enclosed the story of his latest encounter:

'Sockeye is a small humpback whale with an incredibly active spirit, a deformed jaw, and lots of tooth rakes from orca on his flukes. Some of his summer is spent near Cape Cod, and he's been seen for two years in the Caribbean during the winter.

'I saw him first in June 1985, just west of the tip of Cape Cod, as I was trying to beat the worsening weather to port. He was feeding on sand eels

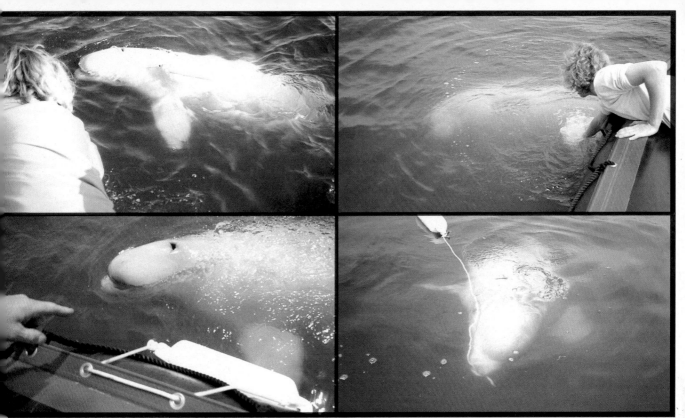

PLATE 37A

Credit: Kevin Vanacore

At New Haven, Connecticut, U.S.A. in 1985, a solitary beluga whale named *B.W.* interacted with people. Here it is towing a fender, hiding it beneath the boat and then observing human responses.

PLATE 37B

Credit: Bill Rossiter

In 1981 *Physty* came to New York Harbour. Thousands thronged the shore as the young sperm whale received medical care and recovered to swim free eleven days later.

PLATE 38A

During a close approach with a humpback whale named *Ibis* off Cape Cod, Bill Rossiter was able to discern the whale sex as she sported around his boat upside down, pectorals splayed, showing the genital hemisphere (below).

PLATE 38B

Credit: *Bill Rossiter*

ll Rossiter has had intensive encounters with *Stub* (above), *Sockeye* (below) and *Trident*.

Credit: *Gillmore — Centre for Coastal Studies*

Credit: *Bill Rossiter*

ent playfully carries a stick, a whale equivalent of the seaweed game.

PLATE 40A

Credit: Bill Ross

On Stellwagen Bank the humpback whale *Catspaw* entertains tourists on the Whale Watch boat *Dolphin IV*.

PLATE 40B

Credit: Bill Ro

In the same area the humpback whale *Crown* breaches from the water. Such behaviour inspires awe in Whale Wa tourists, and increasingly it appears that some performances are for their benefit.

and I stopped to watch because his style is very different. Then I saw his jaw, bent up like a sockeye salmon's, and realised he had to be different to catch fish.

'After a while we ended up moving parallel to each other, and I realised he was playing. For 20 minutes or so I tried to forget the darkening weather and rising chop, keeping the dim flash of his pectoral fins in sight between his surfacing to breathe. Our game became a zigzag, stop and go, slow and speed-up thing that he could have ended at any time, and which I tried to end twice. But he kept challenging me to stay parallel and abeam, no matter what. I was the one who quit, turning back to port for safety's sake, but disgusted at myself.

'Three days later my wife Mia and I were sitting in our little zodiac on Fin Alley, a gently curving line that sweeps in eastward from the open ocean and passes a hundred metres from the very tip of Cape Cod, at Race Point. It is a 20-metre bottom contour that's used by all sorts of marine life, probably for its upwelling and food potential. Its name is obvious to all who sit and watch a single file of finback whales pass into Cape Cod Bay from the ocean, or out again. Half a kilometre or so apart, they stream in. We've used this track to generate lots of photo-IDs and observations of finbacks passing by. Often they interact with us if they have nothing else on their minds. But this day, the 25th, we were watching humpbacks, Sockeye and Speckles, feeding on the same concentration of sand eels.

'Sockeye's style of flicking his tail flukes, swinging his head hard, and sweeping his pectoral fins like wings, helped him feed even with such a badly deformed jaw that I could not tell whether his mouth would open far enough. But it must open adequately because he appears healthy. His small size could be attributed to his youth.

'We observed and documented his moves and bubble-technique for about an hour and a half, noticing some definite patterns and periods which looked like a whale casually throwing himself around. Then, rather suddenly, he moved right at us, slowly, at the surface. We were ready, hoping it might happen. For the next 35 minutes, as *Dolphin V*, the whalewatch boat, moved in to observe from a hundred metres away, Sockeye never really stopped moving, turning to pass under or by us, always so close that some people got concerned about all the white water and flukes or fins flashing around.

'Over the vessel's p.a. system, scientist Dave Matilla explained that the Rossiters do this often, knew what they were doing, and were in no danger — but that no *sane* person should try it! We were never worried because the whale seemed concerned about our safety, but tried to get just as close as he could. Twice he touched the boat lightly, obviously by mistake because he moved off a bit and I'd say he was embarrassed.

'Leaning over the side, hood, mask and snorkel on, I watched in comfortable fascination as he worked to get eye to eye contact. He almost stopped then, and I interacted with him in several ways. We mimicked each other, finding ways to get responses, trying to show we were mutually interested. Sockeye was in total charge. He was disciplined and caring towards us, passing once so close his tightly tucked pectoral fin was all I could see as it filled my visual field with blinding white.

'The turbid water kept visibility to less than five metres, and several times I could only see the approach of a pattern of scars or barnacle circles, trying to find his eye more by its orientation to his mouthline than anything else. For the most part his body was black.

'My Nikonos camera got in the way. Trying to document this kind of event interferes with much that's happening too fast to record, anyway. The camera and all the technology we need to study this kind of moment, gets in the way of the moment itself. Some kind of solution is needed, because at moments like these we have the choice of sharing the event with the whale, or later with people, but not both.

'My photos of Sockeye's eye evoke wonderful but static memories: flukes and fins too close to focus on; scars and marks lost on a giant body; and overall a murky haze from the water. My mind remembers much better than the camera. Those memories flow with the grace of 18 tons or so, pivoting effortlessly to pass eye to eye with total control, and tiny creatures that think the whale is the whole world, and scars caused by, I hope, playful or teasing orca in some distant travel.

'It was easy to lose sight of him, and just to be sure and safe, I would rise up and ask Mia where Sockeye was. She bounced between tears and laughter, as awash in this whale as I was, but from a different perspective. Where I saw only part of him underwater she watched all parts of him splash around close enough to touch. That, by the way, is a rule we don't break. If a whale or a dolphin wants to be touched you'll have no doubt about it. Lifting up to ask and see, I once found Sockeye lifting his head out on the other side of the boat. I remember the question that seemed right to ask; he did seem to be looking at us above and below the water at once.

'There is no way to describe the reality of a humpback that wants to get close up and become personal. I don't construe it as mystical. This was an individual reacting to something interesting in a very individual way. But his exact motive is impossible to assign. I can't even tell with certainty that he knew we were alive separate entities from the boat, or as interesting as we humans think we are. I sensed a mind behind his eye, but what could he tell about mine, behind a plate of plexiglass, both eyes together on a face that might look pretty weird if you think about the perspective involved?

'Much more importantly, what can we learn from events like the opportunity Sockeye gave Mia and me? Are we failing to make each of these events reach its full potential? This isn't criticism, it's a challenge. A marvellous challenge.'

The Rossiters' Approach — Bill and Mia Rossiter are convinced cetaceans possess a level of mental awareness that justifies treating them differently from other life forms:

'Not equal, or greater or lesser than man, but differently. This is one of the best of the many logical reasons for terminating our long history of exploiting them. When a cetacean chooses to approach us we are given a huge opportunity to understand its reality. How many other large-brain, wild creatures deliberately approach humans to interact benignly? As the whale surveys us we can observe and document. From distinguishing marks we discover uniqueness — personality, gender, physiology and life-history. We probe one whale's visual capacity, learn another's lineage, and witness the directed behaviour of a third.

'Most precious are the glimpses of mind. *Cognitive ethology* is the study of behaviour to understand a creature's thought processes. Mia and I attempt to interact and stimulate; we respond and extend. We want to know their thinking and show them ours. According to Heisenberg's Uncertainty Principle, the presence of an observer affects what is being observed. Extending this from particle physics to the behavioural sciences, we can reduce the uncertainty if we become the *reason* for the behaviour.

'On two-thirds of the days we have sighted cetaceans from our inflatable, Mia and I have experienced at least one close approach. Sixteen species of whale and dolphin have been within a metre of us. Some encounters were just single passes; others overwhelmed our senses and endurance.

'In a close approach we work to establish and enhance communication. It is a game with commonsense rules. Both sides display benign curiosity, discipline, respect, trust and even vulnerability. We respond with interest and try to act interestingly ourselves. Scientific discipline must work with open-mindedness; objectivity with intuition. We try to be perceptive at the very fringes of the senses. Even if the source of a stimulus seems within us but is unknown, we react. A simple sound is our acoustic signature: "me". We mimic and build on mimicry with modifiers, such as "correct". We use simple gestures to test, but we do not control, invade, chase or take risks. Everything is aimed at prolonging and intensifying the fragile encounter. It is a positive and productive game.

'From across an enormous gulf, the mind behind that *other* eye joins in a mutual exploration with stimulus, response and recognition. Two species relate for reasons other than survival, sex or social hierarchy. Non-verbal communication tests, limits, and meets certain needs. External events may interfere and if the encounter stops, we can only wait for renewal. We can not pursue.

'There is nothing mystical in all this. But a sense of mysticism has been attached to cetacean study by people impatient for answers, convinced of their own beliefs which are supported by few facts. As a backlash against this mysticism which is taken as a personal affront, cetologists have developed strong biases against cognitive study of cetaceans. Sadly, a young researcher, venturing into this field, faces enough scepticism to damage both reputation and career.

'To discover the reality of cetaceans, we need tools to extend our perception and to document what takes place without impairing our relationship with the whale or dolphin. We are not alone in accepting this challenge but we need help.'

Subsequently, Bill Rossiter has told us of other Stellwagen Bank encounters, such as the cow and calf finback pair that insisted on flanking his inflatable as they all moved along abreast. Despite Bill's reservations about intervention, he had no choice! Another time a 16-metre fin whale made a series of gentle, slow, ultra-close approaches. A rare right whale insisted on lifting his boat in play so that Bill had to fend it off with a hand on its back. Then there was the negative response of a humpback when Bill startled it by broadcasting underwater sounds at a hundred metres range. An angry demonstration was not lost on Bill. There was a hectic day with Pacific whiteside dolphins off Cape Cod, and a young orca that spent a month in nearby Provincetown Harbour, charming the public, playing with boats and accepting gifts of fish.

Ocean Contact: Newfoundland

Further north, near Trinity, Newfoundland, is another summer feeding ground for humpback, fin and minke whales. There, Canadian bioacoustician, Dr Peter Beamish, is engaged in long-term studies which he finances by taking enthusiasts along on his 'Ocean Contact' whale-watching trips, and they assist in his work. As at Stellwagen, the whales have become increasingly curious, and close approaches are a valuable aid to the studies. Peter is now able to correlate a number of humpback sounds with behaviours and venture some interpretation of their meaning.

On 21 June 1979, Dotte Larsen witnessed a humpback response to music while on one of Peter's expeditions. Dotte is a New York cetacean enthusiast with global experience and a vast collection of excellent photo studies. This day their boat was in the vicinity of two whales. Peter Beamish had been playing harmonic intervals on his accordion. When he launched into the tune 'If I were a rich man', Dotte noticed a dramatic change in the whales' behaviour: 'From the rapid, determined movement of what appeared to be concentrated feeding, they suddenly slowed to a measured, almost lethargic motion. In my mind there was no question that the whales were responding to the music, us, perhaps both, but most certainly it was communication and contact. We could all sense it. Just beneath the surface the whales were side by side and barely moving; flippers out-stretched, softly surfacing to blow, then hanging loose as before.'

Still playing his accordion, Peter and Jeff Goodyear got in an inflatable and were able to go within metres of the whales without disturbing them, while Dotte recorded the episode on film.

Peter Beamish has an active programme of rescuing whales entangled in fishing nets. At such times, his team has been able to attach radio transmitter tags and track the whale's subsequent movement by satellite.

On 19 July 1981, Tom Goodman was alone in the water with an entangled humpback, while other people in the vicinity prepared the rescue. After photographing its plight Tom was swimming around it in clear, still water: 'It seemed to be watching me. After several minutes, the whale began to vocalise and then respond to my mimicry of its vocalisation. After my imitation of the previous sound, the whale would repeat the same sound. This continued for about ten minutes until a sudden noise indicated the zodiac was returning to continue the release procedure.' For the next seven days, Theo was tracked by satellite.

On 23 July 1983, Tom and Peter released Theo, the net-entangled male humpback which had previously been fitted with a radio tag. Twenty minutes later the whale surfaced near their inflatable. A rope was still caught in the baleen plates. As the men hauled on the rope, 'Theo at first resisted, then hesitated; then slowly allowed the zodiac to be pulled within half a metre of its left side. The whale rolled, lifted its left eye above water and participated in what seemed like instantaneous visual communication. Opening its mouth it then allowed complete removal of the rope and net.'

On 14 July 1982, two humpbacks were being studied at close range from an inflatable. The larger was recognised as one photographed 100 kilometres away three years earlier by Hal Whitehead. Then the smaller whale, about 11 metres long, swimming slowly in the same direction as the boat, moved its tail sideways and gently lifted the forward half, including its occupants, into the air before lowering it equally slowly. For the next ten minutes the whales hovered around slowly on the surface.

To the south in the Bay of Fundy, Maine, scientist Scott Kraus recorded that a young nine-metre right whale managed to lift their ten-metre boat a couple of inches out of the water and set it down gently, before lying alongside almost motionless.

Boat lifting seems to be a deliberate, communicative whale gesture and incidents have been recorded with grey whales in the Mexican lagoons, where one individual called Lifter specialises in this — but only with inflatables. In Patagonia, researcher Katy Payne experienced it with a right whale. Depending on context it may signify trust, a warning, or play.

On 19 July 1982 in Trinity Bay, English diver Robert Mann met a mother and calf underwater. Two Ocean Contact inflatables had been watching the pair breaching and flipper-slapping for some time. During a lull in activity the expedition decided to have lunch. Close by a whale spy-hopped as if seeking a better look. Robert Mann was eager to meet them underwater. Since they were so close to the boat he rolled in only to be left alone in the ocean as the whales seemed to continue on their way. He snorkelled away from the boat watching a school of capelin darting erratically below. Some were dazed and uncoordinated, swimming upside down in various directions:

'Then, from even deeper, a stream of expanding bubbles trickled past me to the surface. Squinting to find their source, a faint white shape appeared on the limit of visibility. Gradually out of the blue a white flipper formed, then two, and finally four glowing white flippers as the two whales materialised far below.

'The calf slowly ascended towards me, belly up. Flippers extended, it hung motionless only six metres beneath me. It then rolled onto its side to give its curious eye a better view. The mother crept in and gently nudged her calf into an orbit around me. Even by swimming as fast as I could, I was unable to match their speed. When they did slow however, I was allowed to approach quite close, too close in fact, for one whale swung its flukes around to avoid me and then once again kept me at a distance, leaving me in a confusion of bubbles and turbulence.

'With a graceful increase in the sweep of their flukes they vanished back into the depths of the Atlantic. For 15 minutes I had been studied by two whales who were so completely at ease in their world that they offered an intruder a gesture I was unable to reciprocate.'

Caribbean Encounters

Each year humpbacks that spend summer months feeding in areas such as Trinity Bay or Stellwagen move south to winter in the Caribbean. There they have their mating and calving areas: broad shallow sand banks around certain West Indian islands, especially the Silver and Navidad Banks.

Ever since 1973, scientists have observed and photo-identified the humpbacks in this area, and compared results with similar studies in the north. In their excellent book *Wings in the Sea*, Professor Howard Winn and his wife Lois describe techniques for observing whales underwater: 'Diving among whales without disturbing them is difficult. We soon learned the best way to observe mothers and calves is to stay in one place and let them come to us. Calves were allowed to come quite close to a diver or ship as long as no movement was made *toward* them.'

The scientists would watch until the whales were within 200 metres of the ship before descending on scuba, in the warm shallow water and waiting as the whales approached: 'Underwater humpbacks appear slim and agile as they move with their slow grace. Their flukes seem to flow, the long flippers (six metres across) undulate slowly and their bodies are supple with great muscles rippling beneath glassy black skin. They look at us with eyes that seem incredibly small. We are awed by the physical power and presence of such huge creatures. As they turn to avoid us, we see how efficiently they use their long fins for turning and balancing . . . As the humpback accelerates the fins are swept back like the wings of a jet.'

Near the Florida coast is a small group of islands called the Dry Tortugas. On 25 July 1976, oceanographer Dr James W. Porter came upon a pod of 30 *Pseudorca* resting on the lee shore of one of the sand cays, facing towards the shore. The whales' backs were awash in the clear shallows: 17 females and 13 males. At the centre, the largest — a six-metre male — was sick. Blood seeped from his right ear. The other whales were not stranded and periodically would shift towards the middle of the wedge-shaped formation.

Nearby, Dr Porter entered the water, intending to snorkel amongst them. The outermost whale headed over, lowered its head and slid beneath him. Slowly its body rose, lifting him clear of the water and bearing him towards the beach where it slowly submerged. The whale circled him and approached again. James Porter dropped his snorkel on the verge of calling for help, but at the release of his snorkel the whale seemed to lose interest, veering off to rejoin the group. The pattern was repeated three more times.

At that, the scientist walked along the beach to re-enter the water on the other side of the group. The flanking female responded in the same way an equal number of times.

The *Pseudorca* accepted his presence to take intimate underwater photos of them when not using the snorkel. James Porter wondered whether it sounded like a clogged blowhole. He recalled tales of drowning people being pushed ashore by dolphins and small whales. After three days, the large male died and the rest of the group returned to sea.

South American Right Whales

In Argentina, off the windswept Valdes Peninsula, Patagonia, there is a nursery of southern right whales, *Balaena glacialis australis*. Each year these cetaceans migrate from their Antarctic feeding range to two large, sheltered bays. There, in fairly shallow water, they breed, give birth and nurse their young.

Argentine diver, Dr Ricardo Mandojana, has dived intensively with the Patagonian right whales. He recounts some intimate, gentle meetings, and some rather severe warnings when the whales considered him intrusive or infringing in a way they considered tabu.

In spring 1975, Ricardo had his first good opportunity to swim with a right whale and photograph it extensively. A few metres from his inflatable the 14-metre female floated motionless, just a small portion of her smooth, finless back exposed, the massive black head with its knobby callosities just visible in the greenish water. Ricardo eased himself into the chill, six-metre clear Patagonian sea and swam towards a vague shadow heading slowly towards him. At three-metre range he started taking wide-angle pictures. A few metres off, the whale veered. In a slow, co-ordinated movement the pectoral fin dipped; the colossal body arched and the head turned laterally. Ricardo dived below it for a new angle, and to give the whale a better view of himself. Her body scarcely moving, the whale's eye followed his manoeuvres.

'As I neared one broad, rectangular, paddle-like flipper the whale carefully retracted the limb to avoid hitting me. I dived to the top of her head, took a couple of pictures, carefully swam to the "nose" and dropped across the whale's path for some head shots. She was motionless. I surfaced for air. Below, the whale was "standing" almost vertically, her head a few feet from my fins, looking at me. I was intrigued to see how profusely her skin was peeling; flakes up to a metre long were hanging from most of the body surface.

'We had been swimming together for more than half an hour when I decided to attempt direct contact with her during my next plunge — to touch and caress her skin. Finning my way down, I got to the roundest part of the belly and took pictures of the large, irregular piebald markings (which are an additional means for identification of individuals) as well as the head callosities. I swam around the rotund body, over its dorsum towards the head. Again she glued her pectoral fins to the body and swerved delicately to get a better view of me. The eyes of the right whale are set well back in relation to the "nose", almost completely laterally and above the corner of the mouth. A lateral view gives a wider angle of vision.

'I stroked her side and back gently with my bare hands. The skin felt smooth like slippery rubber. I was surprised to feel a distinct quiver, like the tremors a hand elicits from the skin of a horse when patted near the muzzle. I realised how extremely sensitive the cutaneous surface of this gigantic creature was. My fingers, in relation to its immensity, would compare to the legs of a fly standing on a human body. With difficulty, I managed to remove one of the cyamid lice for later identification. Abundant on the head, especially in the callosities, and every fold of skin, I wondered if they were uncomfortable for the host, considering the skin's sensitivity?

'Now the whale swam slowly and steadily while I paused to admire its magnificence. I moved slightly from its course, avoiding the wake of the six-metre plus tail. The twin flukes moved sinuously up and down to propel the powerful body. My viewfinder was filled by the gigantic appendage. As the tail dropped from the surface, a curtain of silvery bubbles followed its massive weight. I shot my last frame. It was late and cold; the whale melted into the ocean gloom. Tired but euphoric, I climbed back into the boat.'

On an expedition in October 1980, Ricardo met another very tolerant whale. After taking photos, he played with it on the surface for a while. Then he started to touch, very gently, her lower jaw and upper lip:

'I could feel the short thick hairs growing on the polished skin surface. The whale started to raise her enormous head more and more above the surface. She could not be standing on the bottom as the water was deep. More than a third of her body was vertical, above the surface. I stroked her ventral "neck" which was smooth, without pleats. She moved gently, like dancing in slow tempo. She was responding to my caresses with visible pleasure.'

On this same expedition, three divers, Drs Vicky and Ricardo Bastida, Argentine marine biologists, and Ricardo Mandojana, were diving with an extremely cooperative 15-metre female, when the unexpected happened:

'Ricardo and Vicky were close to the head and I was approaching them, swimming over the whale's back. I dived and caressed its side a couple of times (we had been doing this for a while now) and swam toward the head. My right hand was holding the camera while the left arm was stretched forward, close to the whale's body. When my left hand was near her eye, the whale was suddenly startled. I now deduce she was looking forward, towards the divers in the front, and did not expect the sudden appearance of "that light pinkish object" (my bare hand), so close to her right eye.

'The whale made a rapid frontal take-off, thrashing the head as well as its tail sideways. I was able to avoid the flukes but Vicky was hit with a head-on butt on her hip. Fortunately, she got safely back to the boat and the thick neoprene of her wetsuit cushioned most of the impact. Only a mild bruise remained for a few days. I believe that the butting was *not* totally intentional but this is speculation because we know of other occasions where violent lateral and up and down head movements were part of aggressive display, without a subsequent attack (Bill Curtsinger). Perhaps Vicky was just too close.

'Once we were all safely back in our boats, the whale stayed around slapping with her tail for a while. Then she came very close, turning slightly sideways and her blowholes shot a few well-aimed jets of water at us.'

In this same area Dr Roger Payne and his wife Katy have done many years of study. One one occasion, he recounts, Katy was watching from a small boat. Nearby were mothers and calves: 'A whale approached and circled the boat closely, almost touching it several times with its head. Then it turned its flukes to the skiff, swished them strongly from side to side, backed up and placed them under the stern, raising the whole boat, passengers and all, about two metres in the air. The whale let the captive craft hang there for a long minute, then lowered its flukes slowly without tipping the boat. Over the next five minutes the occupants made no attempt to free the skiff, and

Credit: Bill Rossiter

rom his base in Newfoundland, Canada, scientist Peter Beamish interacts intensively with humpback whales. Here one iows interest as he plays the accordion.

Credit: Bill Rossiter

f Race Point, Cape Cod, huge sleek finback whales encounter Bill Rossiter. The second largest whale species (22 metres gth, 50 tons weight) and usually elusive and aloof, fin whales can be sensitive, friendly and playful towards humans.

PLATE 42A

Nog, an eight-metre new-born right whale meets Bill Rossiter off Cape Cod — 'a curious, bus-sized puppy'.

PLATE 42B

Right whales, an endangered species, feed close to the ports of American whalers that almost exterminated them. Wh; encounter is now an extremely viable tourist industry.

PLATE 43A

Credit: Bill Rossiter

The friendly grey whales of St Ignacio lagoon, Mexico, have been a major tourist attraction since 1977. *Amazing Grace* responds to human voice and initiates play and touch with the Rossiters.

PLATE 43B

Credit: Dotte Larsen

Few who visit St Ignacio lagoon fail to touch a grey whale as mothers and juveniles throng the tourist skiffs. The area has long been a whale sanctuary.

PLATE 44A   *Credit: Bill Rossit*

An orca pod is a close and complex family group that hunts, travels, nests and plays cooperatively, communicating b
sound, sight and touch. This pod is passing through Johnstone Strait, British Columbia.

PLATE 44B   *Credit: Bill Rossiter*

A male orca named *J1*, about eight metres long, encounters
a familiar research boat in Johnstone Strait.

PLATE 44C   *Credit: Bill Ros*

This solitary orca *Elsa* spent a month approaching peo
in Provincetown Harbour, Cape Cod in 1982.

the whale lifted and set it down two more times with utmost control and deliberation.'

In other accounts warning displays have been provoked by divers getting too close to mating whales. An Italian diver suffered four fractured ribs from the flukes of a male as he tried to photograph a nearby female. Another diver's boat got too close to a love-making pair: the male swam around it hitting the water with his fins but never touching the boat.

Clearly, these whales have shown superb restraint towards humans and do not deliberately seek to injure us. It is equally clear, however, that we must exercise similar restraint, never sneaking up on them to snatch photos; always showing respect for mothers with calves, and copulating couples.

Encounters with Grey Whales of Mexico

In all the world, the most frequent contact between humans and whales undoubtedly occurs in San Ignacio Lagoon, Mexico. Tour operators can now be assured that virtually every passenger will be able to stroke a whale.

Each northern winter, California grey whales, *Eschrichtius robustus*, migrate south along the Pacific coast from the Bering Sea to nursery areas in the warm Baha lagoons. By late December the 35-ton cetaceans are beginning to enter San Ignacio and Scammons lagoons, and Magdalena Bay, where they will court and calve and nurse their young until departure for northern feeding grounds in May — a 16,000-kilometre annual odyssey from Alaska to Mexico and return.

Severely depleted by hunting, the Baha greys received protection in 1947, but for years they were avoided and feared for their habit of inflicting violent attacks on pursuing vessels — especially mothers and calves. Mexican fishermen called them 'devilfish'.

Then, in the early 1960s, there were stories of grey whales lingering around skiffs and tour boats, but they were not trusted to come close. In the winter of 1975, with the growth of US-based whale tourism, came the first reports of curious whales following tour boats for hours, circling and rubbing on the hull. Dr Raymond Gilmore tells of a female presenting herself alongside to allow passengers to pat her head. Dr Bruce Cauble had three females with a calf accompany his drifting skiff for an hour with all four repeatedly surfacing to be touched.

Each year interaction increased in frequency and quality. During the 1977 season, researcher Steve Schwartz found a change in the whales' attitude to scientists. Initially they showed little attention to observers of their daily rituals but soon individual whales would remain with them for an hour or more. By then, up to 1000 people were visiting the lagoon each season and most would have witnessed human/whale contact.

One particular female, named Amazing Grace, delighted whale-watchers for three consecutive years with playful, innovative behaviour such as submerging her blowholes to spray people; spy-hopping beside the boat to look at occupants; accurately wetting them with sweeps of her tail; and imitating the gurgling sounds of an idling motor with her blowholes just below the surface. If the motor was revved she would release blasts of air, just as had been observed among small groups of whales.

Bill and Mia Rossiter were members of a three-boat expedition in 1977, when first contact with Amazing Grace was recorded. During several weeks in the lagoon, whales had occasionally ventured within two metres of their skiffs but there had been no close interaction until late one afternoon when an adult came up beneath them and surfaced alongside. To interest the whale and project their confidence, the Rossiters spontaneously began to sing the hymn 'Amazing Grace'. What followed was historic — the whale began to lift their boat on its head.

'She was very gentle with us. Then she moved about softly, tucking her flippers in close, rubbing and pushing us, and turning repeatedly. Her enormous tail never came near. She was positively cautious. We could easily have been overturned. Even her breathing was restrained near us.

'Vertical alongside, her rostrum just above the surface, we began to touch her. We crooned "Amazing Grace" as we rubbed the smooth skin and felt a gentle return. A flipper was presented, motionless and dripping, inches away. We pulled it. The whale seemed to come closer, as if we had really moved her. Then the incredible mouth opened and we rubbed gums and baleen — yellowish, dripping and wonderfully close. "Amazing Grace", as my wife decided to call her, presented the entire forward third of her body to us, purposefully.

'With mask and underwater camera, I hung over the side and peered into the sun-rayed green-brown haze of the lagoon. Below us she rose like a grey cloud. Then she rolled upside down beneath the boat, her pectoral fins straddling us. With absolute grace and control, she lifted us on her chest between her massive flippers. Then she came up again from behind me, her rostrum gliding within reach. Gently she rolled on her side and paused. We were looking at each other.

'Another time, I placed my hand on her rostrum and pushed — and was pushed back. Again and again she returned at all possible angles and attitudes. Once she paused as her blowhole came by and released a frothy mass of bubbles that boiled around the boat. We had seen whales do this with each other. It certainly seemed to say something. This whale did everything possible to relate, probably testing us in more ways than we knew. Was she toying with us or was she just curious? It all seemed beyond mere play; her concern and caution, her bubbles and noises.

'The session ended when I entered the water. Perhaps she was startled. Immediately she went over to the other two boats watching from a short distance. Now we watched as she lifted and probed them, presenting her body to be rubbed and pulled. They reacted as we had. After 35 minutes, she left.'

That same year (1977), marine ecologists, Steve Schwartz and Mike Bursk, were motoring across the morning-calm lagoon when they sighted three whales leaping acrobatically. They investigated. At close range they observed a courting ritual, the whales seemingly oblivious to their presence. Suddenly the whales vanished — but not for long. The female surfaced ten metres from their boat, circled it slowly and disappeared.

'Moments later,' wrote Steve, 'she is directly beneath us. Imagine our excitement watching a huge form rise through the murky water and gently nudge our boat. The two males return and seem eager to participate. They too shove the boat, ever so gently. A whale surfaces vertically, and with a graceful motion, rests its head on a pontoon. Mike touches the rostrum and the whale abruptly backs away. But after the initial contact the whales grow bolder. They continually keep their upper and lower jaws accessible for scratching and massaging, and we are willing to accommodate. We completely surrender to this astonishing phenomenon.' (*Whalewatcher*, Vol 13, No 1, 1979.)

Year by year friendliness intensified. By 1980 researcher Jim Hudnall was able to film a situation where *eight* whales were circling a tourist skiff: mothers with calves interacting with people. When the time came for the skiff to return to the mother-ship for more tourists, the whales seemed to block its path

227

repeatedly. Each time the engine was put in gear a whale would surface at the bow. When the skipper eventually managed to slip through the gap, Jim believed the whales had deliberately relented and had stopped their teasing. By then some of the passengers were becoming a little distressed!

Whale photographer, Dotte Larsen had visited the whales in 1976, but didn't find them particularly friendly. When she returned in 1981, however, it was a very different story. Their five-metre skiff was cruising slowly over the lagoon with six aboard when suddenly a mother and calf surfaced alongside.

'After checking us out visually, all but thrusting their huge heads into the boat and on our laps, they gently rubbed the boat, rocking us slowly. They pushed the skiff like a kitten with a ball, rolling over and over on top of one another. Then they wanted to be petted, vying for attention, splashing us with their tails or blowing jets of water over us. I stroked them on the head, mouth, chin and back, but it was the eye — that four-inch, unblinking, benign eye — that penetrated my very being. It seemed to speak to me of all the pain inflicted by my ancestors . . . I felt shame and guilt on behalf of mankind.'

Diving with Grey Whales — The murky water of San Ignacio Lagoon make underwater observation of the grey whales difficult, but cameramen Marty Snyderman and Howard Hall were commissioned to get footage of them, covering every aspect of activity. With diving permits from the Mexican Government they set about their project in 1981. At first they admit they were too pushy, spurred on by the intrusive demands of their filming schedule.

Sighting a group of three courting whales, they swam towards them. Howard saw an adult heading his way, so dived to ten metres, camera ready, and waited. In the gloom, huge shadows swirled about. Suddenly, at two metres range, he was looking into an eye surrounded by a mountain of grey flesh. He was certain he saw fear in that eye. As it accelerated off, a tremor passed through the whale's body. Two seconds later, there was a fleeting image of great flukes smashing into his body, and he lost consciousness.

Howard received two broken ribs and a fractured arm. His camera, left running, captured frames of a four-metre pink organ. He believes the whale, in the midst of courtship, had not sensed his presence until they met eye to eye. 'Certainly I cannot blame the whale. I had no business approaching the animal in this manner.' (*Skindiver*, September 1987).

From their mistakes the divers learnt it was best to let the whales come to them. Later, they had wonderful experiences. On snorkel, Howard held the side of the boat and watched for more than an hour as three whales rolled over and over on their sides as though dancing. One eye would roll up within his reach, and then the other. With each roll, the pectoral fin missed his chest by inches. The game continued as long as their outboard was running in neutral. While the adult trio rolled around the boat, calves would spin vertically, heads out and close to the hull, at nearly 60 revs per minute.

On another occasion, Marty Snyderman was approached by a cow and calf, and for 30 minutes he free-dived and filmed them. The calf was first to show curiosity, coming within inches. Then the mother seemed to relax and come close too. Once she exhaled below him. 'The stream of bubbles knocked me for a loop. Fortunately, neither animal surfaced directly beneath me; both seemed acutely aware of my exact location. Their movements were incredibly precise and their body language excellent.'

Marty watched as they gently touched, rolled against one another, rubbing

228

'faces' and spinning. Twice the mother fluked forward at great speed. The backwash dislodged his mask. She brought one fluke within inches of his head, but never made contact. When his fingers touched the whales, they seemed sensitive to his contact. He became so confident, he filmed close to the calf's eye but there was a price to pay: 'It proceeded to bounce me on its nose four times.' (*Skindiver*, January 1982.)

In their five-year study of grey whales, Mary Lou Jones and Steve Schwartz evaluated the impact of whale tourism on the lagoon population and concluded that it does not seriously threaten the whales but should be closely monitored for the future. They noted that it was in the interest of tour operators to show deference to the whales and that the increasing incidence of friendly behaviour suggests the whales have become accustomed to tour boats.★

Following a superb article in *National Geographic* magazine by these biologists entitled 'Grey whales play in Baja's San Ignacio Lagoon' (June 1987), a reader Ken Stanley wrote to the editor: 'I observed that the cows often supported their calves high in the water as though to give them a better vantage to see us. They repeated this manoeuvre so close to our skiff that I could stroke the muzzle of a calf with one hand and that of its mother with the other. I know of no parallel situation in which a mother in the wild intentionally presents her offspring to see, be seen and even touched by an alien species.'

Not surprisingly, friendly grey whale behaviour has now spread beyond the Mexican lagoons and there have been reports of close approaches along the Californian coast.

In September 1982, Jim Hudnall and Ann Spurgeon documented an episode off Vancouver Island, when one of a pod of four approached several boats and rubbed on them. On three consecutive days a whale came to Jim's inflatable, let them rub its mouth, rolled on its back to be stroked and lifted their boat very gently.

★ Chapter 14, *The Grey Whale*. Jones, Schwartz and Leatherwood. Academic Press, 1984.

Sea of Cortez Encounters: Fin Whale

The fin whale, *Balaenoptera physalus*, is a sleek, fast giant. Exceeded only by the blue whale in size, it grows to around 22 metres and can maintain more than 30 kph for long periods. Reputed to show little interest in boats, this is clearly changing in areas where it is protected.

On 20 November 1982, Kenneth Bondy and two friends were fishing in the Sea of Cortez, 650 kilometres south of Mexicali. Sixteen metres from their orange inflatable, a large fin whale blew. They hauled in their lines. Two minutes later it blew again, three metres away. With the third blow it was heading straight for them. Gracefully it glided beneath their hull, pausing with a slight roll to view the object above. They estimated its length between 16 and 20 metres.

Then the whale began spy-hopping close by, rising vertically until the eye was just out of the water and the tip of the rostrum two metres above. These postures were held for several seconds, with slight movements of body and flukes to maintain position. The contact lasted over 40 minutes with the whale showing every two to three minutes. It was visible to the men for the entire period, circling below them and manoeuvring between surfacings. 'We were spellbound by this amazing encounter and the obvious communication with this enormous, gentle creature.'

The episode ended when another boat motored over, thinking they were in trouble. The whale surfaced ten metres away and left.

Orca of the Pacific Northwest America

Cameraman Bob Talbot shuddered as the 49-degree water of Johnstone Strait penetrated his wetsuit. Companions in the inflatable told him two orca were headed his way — a cow and calf. He swam through kelp to a clearing. Suddenly, he was face to face with the calf. As he filmed, it was joined by the mother, and the pair swam to and fro in front of him. After the third pass the mother approached him directly, before they swam off.

These antics were repeated several times. Then a male orca circled the boat, vocalising and blowing bubbles. Bob re-entered and soon saw swift-moving white patches through the green haze. The whale spiralled towards him, closer with each pass. When his film ran out Bob decided to return to the boat, but the whale intervened. 'It was obvious he was not yet ready for me to leave.'

Bob accepted this and the circling resumed. At times the orca passed close enough to touch. Every second or third pass it would pause, stop vocalising and hang motionless, looking at him.

'I was filled with ecstacy and apprehension, feeling great joy at being the centre of the whale's interest, but also feeling trapped and out of control of the situation. My apprehension came not from any apparent aggression, but from possible play that might get out of hand.'

Circling continued for several minutes until the whale settled for one final pause. 'He then swam gently, quite close to me, emitting a parting vocalisation and, with one sweeping motion, disappeared in the green water.' (*Whalewatcher* Vol 15, No 1, 1980.)

Orca Studies — At all four corners of North America, human/cetacean 'hot spots' occur: great whales off Cape Cod; dolphins in the Bahamas; grey whales in Mexico; and *Orcinus orca* in Washington State and British Columbia.

In the Johnstone Strait area, at the northern end of Vancouver Island, a maze of waterways and salmon-rich territory supports a resident community of some 200 orca. Since 1973, intensive field studies and photo identification has revealed an astonishing picture of orca social life. Every individual, in four distinct groups, has been identified and followed for 14 years. Because their residential lifestyle makes them especially accessible, this orca population is *probably the most thoroughly studied community of wild animals in the world*.

When a big male orca with his towering fin is seen among a group of smaller ones, it is usually assumed that he is 'big daddy with his harem'. This is totally wrong where Canadian orca are concerned. They live in stable family groups called pods, each with its own dialect of distinct pulsed calls. In the entire study period, no orca has entered or left a pod except by birth or death. Females live to around 100 years and males to around 50. Like the long-lived elephant, each orca pod is based around a dominant female — a matriarch with her sons, her daughters and their progeny. Even mature bulls stay all their lives with mother, and often 'baby sit' juveniles. The birthrate is low — about one calf every ten years.

When orca pods meet, a strange ritual takes place called 'intermingling'. While this may lead to mating, it must have other important social functions.

The pods form two tight lines and approach head-on. At close range (10-12m), they hover motionless at the surface. Then the groups submerge and intermingle, moving *slowly* in tight, milling clumps, many touching, rolling and brushing each other. Pairs and trios spy-hop, but activity is splashless and relaxed and may continue for over three hours. Towards the end, momentum increases with noisy leapings that merge into a session of pure play. Thirty per cent of the time, during intermingling, the whales are silent, which is most remarkable for these vocal creatures.★

From the world's first orca capture for live display in 1964 until legal restriction in 1976, the accessibility of the northwest orca population led to intensive exploitation. Solely for public exhibition, whole pods were entrapped and 64 were 'cropped'. A 1970 study revealed that a quarter of captured orca had bullet wounds — conflict with salmon fishermen was intensive. World trade in orca as aquarium exhibits generated the need for studies of them in the wild.

Paul Spong — In June 1970, Dr Paul Spong, a New Zealand biologist, began to observe the Johnstone Strait orca and set up his now famous fieldstation on Hanson Island. In a sensitive account, 'The Whale Show', Paul tells how his orthodox approach was discarded when he found his kayak was accepted and he could move among the orca for hours at a time, day after day.★★

On one occasion, Paul was enshrouded in thick fog with orca breathing on all sides. As he alternately played his flute and paddled, the orca stayed with him and he was induced to move along in their company. One group swam continuously on the surface, four either side of the kayak. For ten hours Paul went with orca, regardless of his direction, until an island loomed out of the fog and he landed. So began his long-term interactive studies, still in progress 18 years later.

Eric Hoyt — In 1973, Eric Hoyt began a special project to gain acceptance by the orca, and to film their living patterns. His superb book, *The Whale Called Killer*, documents the adventures of his team over seven summers as they gradually came to know and understand Stubs pod.

Initial contacts were musical. In July of that year, Eric played a synthesiser to orca near his hydrophone/speaker gear. His imitation three-note whale phrase was met by a chorus response — several whales imitated his effort in harmony. 'They did not repeat their own sound; rather they duplicated my human accent.' After he repeated the phrase four times, two whales answered. Again they mimicked his slow, stilted attempt at whale sound, confirming the initial response.

Hoyt's book narrates each session's interactions as the whales became increasingly tolerant of human company. From brief boat encounters and the fleeting scrutiny of divers, the orca eventually allowed the team to observe gamesplay, group resting and intimate courtship rituals. Juveniles would play tag with strands of kelp right alongside their inflatable. Off 'Rubbing Rock' beach, ten orca swam in a tight circle, vocalising strangely. Then they broke off and submerged.

★'Behavioural Biology of Killer Whales' — Richard Osborne 1986.

★★ *Mind in the Waters*, editor J. McIntyre, Scribner, 1974.

In the shallows they were rubbing their bodies on the smooth stones, rolling over and over, tangling in the kelp, touching and mouthing each other. Then, on a rare occasion, the team saw a pair slide against each other, belly to belly, and mate.

In 1977, Hoyt's group were able to film interplay between divers and orca, and to swim with them through the Strait. 'It was the kind of scenario we could not have imagined in previous years.'

By 1979, the orca were staying so close around five scuba divers that the team began to worry. Would people take advantage of such trust? If the whales became too friendly would they get shot at? The orca showed a special affinity for human revelry, assembling out in the dark when camp parties were in progress. Human intermingling, perhaps?

Jim Nollman — Increasingly each summer, a flotilla of small craft follow the whales as they traverse the Strait — whale-watchers, researchers, film crews, all exerting such a heavy pressure on the orca there is concern they may suffer stress.

But Jim Nollman's passive approach is hard to fault. Each August since 1977, he awaits the whales at night with a growing assembly of musicians and the whales interact with his underwater sound array.

In 1979, Nollman's guitar mimicked an orca phrase. Then, he claims, the orca broke that phrase down into three parts. When his guitar erred in repeating it, the orca slowly rehearsed the phrase again. When the musician corrected himself, the orca responded with a new, upswept note at the end.

In 1986, Nollman wrote: 'The very best night was when we had a Tibetan Lama chanting. Ten orca were within 50 metres of the boat. One came close to the underwater speaker. The whales were silent. When the chanting ceased they began whistling and grunting loudly. Then the lama was joined by violin, guitar and electric organ. The whales vocalised with us, sliding their notes up and down scale to meet ours. It is becoming clear that the orca relate more to the level of the human musical bond than to any specific tune.'

Friendly Whales in Hawaii

Snorkelling quietly, Jim Hudnall could hear a humpback singing somewhere below him in the depths off the west coast of Maui. A few minutes earlier, the researcher had seen the whale raise its flukes and vanish. He knew humpback songsters spend from 18 to 22 minutes below, but no human had ever seen one performing. Peering to and fro, he hunted for a glimpse of the whale as the sounds grew louder. Then a trace of something white appeared in the blue abyss.

Jim switched from snorkel to the tiny scuba tank he uses for whale study. 'As I descend I can clearly see the outstretched flippers of a motionless humpback. I pause a few feet above its huge flukes. The whale sings a note that is shorter than usual, then turns to look at me. The song ceases, and the whale approaches until we are eye to eye. A slow roll and its genital slit faces me. Are we to "rub bellies" as dolphins do? No longer do I wonder what sex this humpback is: very definitely a male. Finding me unresponsive to his first advance he glides off, turns around and repeats the performance. I ignored him again. He slips away into the near distance and resumes his song.'

Each winter, humpback whales of the north Pacific migrate from cold Bering Sea and Alaskan feeding grounds to the warm waters of Hawaii and other tropical islands, where they mate, calve and rear their young. This population of humpbacks is only a remnant of a tribe which in 1905 numbered about 15,000. Now only about 2000 survive.

In early November, whales begin entering the sheltered Four Island area, adjacent to the island of Maui, and by mid January significant numbers are present. The population peaks in mid February and remains relatively high until mid March. By the end of June they have virtually all withdrawn. This humpback nursery is close to high-density human population centres, and intensive tourism.*

When Jim Hudnall first came to Maui in 1974, not much was known about the humpback nursery, the singing whales and their living patterns. Jim returned in 1975, and commenced his study using the simplest of equipment — a small inflatable boat, snorkel and his lightweight mini-scuba rig. Jim often operates alone leaving his boat drifting on a sea anchor while he observes and films the whales.

In 1975, he began recording their songs with a hydrophone. This began when he found himself within audible range of a singer. He and scientist Jim Darling made snorkel sounds in reply, and suspected that the whale may have mimicked them. Hudnall obtained recording gear to document the songs. From that time, he never again ventured a communicative approach, believing it was more important to document the unmodified songs initially.

In his first year of study, Hudnall identified individual whales from the

* Pressure on Hawaiian humpbacks has necessitated protective laws: today divers must have special permits to approach them.

234

pattern on the underside of the tail flukes. By 1976, he had realised that this technique was ineffective for identifying females with young. Nursing mothers don't dive deeply, showing their tail flukes as their young are incapable of such dives, and would be left behind. Jim then discovered the upper surface of the pectoral fin was a distinctive and comparatively easy feature to photograph. With assistants, he began a programme of cliff-top observations, and was able to identify increasing numbers of whales, observe births, and maintain a continuous watch of cow/calf activity.

By 1977, whale-watching was in full swing with many observers working in the field. Whale tourism was flourishing. In 1979, Jim established the Maui Whale Research Institute which has a photographic catalogue of individual whales available to researchers.

While with Jim on Maui, I was able to see his films, hear his tapes of whale song and compare notes with him on interlock. Approaching whales, Jim shuns the earlier techniques of leaping in front of them from fast boats for a fleeting glimpse, or confusing them by racing around in tight circles. As long as he moves gently towards them over the surface, the whales show no fear and approach him curiously. His longest encounters have been with humpback cows and their calves. Baby whales are playful and curious to explore their environment. Often a calf would lead its mother to investigate Jim, coming so close at times he expected to be touched — but at the last moment contact was avoided. Even young calves had this precise body control.

Jim remarked on the special friendship that had developed with a whale they call Notchy, after the deep notches on its fins. For three consecutive years he met this same whale. Gradually, it stayed around him longer and longer. He thought it was sick when it came so close he was able to touch it, but days later he saw Notchy with another whale, leaping and racing about in perfect health.

In 1975, Jim had introduced biologist Debbie Glockner to the whales, and she went on to become expert at gaining acceptance through an understanding of their body language. As a result, she and her husband Mark were able to develop a new method of sexing humpbacks by swimming beneath them to observe the hemispherical lobe that females have in front of the genital slit. Then it was discovered that the adult, which usually escorts a mother and calf, is a *male*. Even more exciting, they found that Daisy had a *new* calf with her for three consecutive years, the first humpback known to reproduce annually which indicates that this species may be attempting a recovery. Another whale, Big Hunk, was followed for 11 years.

By 1983, nearly a thousand Hawaiian humpbacks had been identified, and social patterns were emerging. Researchers began to appreciate that each whale had its own character or 'personality': brave, curious, shy or indifferent towards humans. Jim Darling felt that a whale's behaviour towards people probably depended on past experiences, whether they had been good or bad.

'We have learned that whales are not just a herd of big clones. When we kill them we are killing individuals, each one different from all others in its character and role in the herd. Each needless death is a loss to the population forever.' (*Whalewatcher*, Vol 17, No 2, 1983.)

In deep ocean waters off Hawaii's Kona coast, biologist Dan McSweeney and cameraman Al Giddings met a group of about 15 rare pygmy orca, *Feresa attenuata*. For four hours they swam with the tiny whales (2-3 metres) and filmed them. One was so friendly, passing within an arm's length, that Dan gently touched its tail. The whale torpedoed down to 16 metres and stopped, shaking itself like a dog. Then it rejoined the group. Half an hour later the same whale calmly presented itself to Dan, *allowing* him to stroke it.

On another occasion in this area, Dan and others watched as *Pseudorca* preyed cooperatively on mahimahi. Then one came and gave Dan a fish. Towards the end of the feeding frenzy a whale came back and took the fish Dan was holding.

PLATE 45A   *Credit: Gary Dods*   PLATE 45B   *Credit: Gary Dods*

ear the Aldermen Islands, New Zealand, in February 1979 three orca approached the launch *Hirawanu*. A large female
sted her head on the stern platform for several minutes while all aboard stroked her.

ATE 45C   *Credit: Gary Dods*

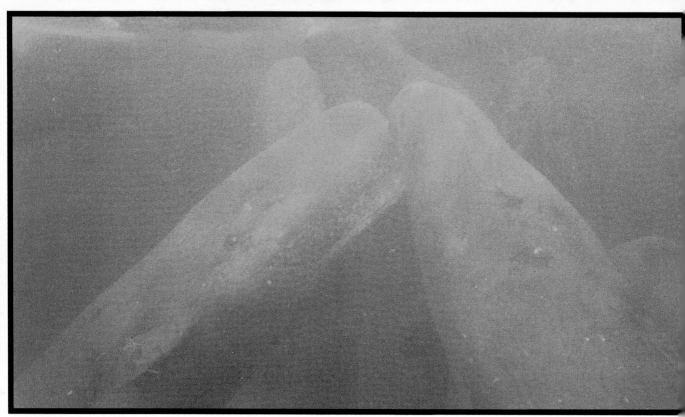

PLATE 46A   *Credit: Rupert Gi*

At Lord Howe Island, Tasman Sea, in April 1987 some sixteen sperm whales hovered in 20-metre-deep waters ne
the reef edge.

PLATE 46B   *Credit: Rupert C*

A strange social ritual in progress as about half of the group lie on the surface, their heads touching. Others are uprig
tails just clear of the bottom.

eeting only gentle acceptance and silence, a trio of divers witnessed this rare ritual for four hours until dusk. The ales showed no annoyance when accidentally bumped by the divers.

PLATE 48A   *Credit: Irvin Rockm*

A minke whale meets divers near Lizard Island, Great Barrier Reef, Australia. Targeted for Japanese 'scientific' whalin
this individual may now be dead.

PLATE 48B   *Credit: Barry Ha*

Off the Kaikoura coast, New Zealand, in spring 1980 this right whale approached the fishing vessel *Pirimai*. When th
men entered the water, the whale arched, lifting Peter Kelliher, and another crewman was raised gently on its tail.

Sperm Whales Mid-Ocean

Sperm whales generally inhabit deep waters far from shore but that has not prevented Dr Hal Whitehead from studying them. For three seasons (1982-84) his ten-metre sloop, *Tulip*, followed them in the Indian Ocean off Sri Lanka. Then he continued in the Pacific, spending two-and-a-half months in 1985 tracking sperms off the Galapagos Islands with *Elendil*, a similar vessel.

Because sperm whales produce regular echolocation clicks as they forage for squid some 500 metres below, Hal found it possible to track their slow progress using a directional hydrophone which could give a bearing on clicks from about eight kilometres range. Sometimes the clicks formed tight patterns called *codas* which might be exchanged between whales, with one repeating its own coda, alternating with codas from another whale. Interpreting these sound patterns is a major challenge for researchers.

In the Indian Ocean, along with a group of about 15 whales, the *Tulip* crew witnessed a birth.* Lindy Weilgart jumped in quickly. 'I could see the umbilicus attached to the calf and the afterbirth protruding from the mother. I was astonished that the calf had bright blue eyes.'

It swam around Lindy within touching distance while the mother hung back watching. Later the mother left her calf in the care of *Tulip* as babysitter, while she dived deep.

Hal Whitehead found that once every day or two, especially mid-afternoon, the whales would cease their deep diving and congregate on the surface for a restful interlude of an hour or so. Slowly they sidled up to one another, to form still, silent, raft-like subgroups. During these social periods the crew would snorkel with them.

'We saw them turn gracefully to watch us; gently stroke one another with their small flippers or nuzzle a smooth, bulbous brow against a vast, wrinkled flank.' (*Natural History* magazine, 6/86.)

As *Elendil* drifted within 30 metres of a female, she turned and headed directly towards them. The other 11 whales in the group turned and began a steady approach. The lead female glided slowly, with no apparent tail movement, to within a few metres of their stern. Next came a huge male. It passed beneath them, a metre below the keel. 'For a few interminable seconds the enormous animal seemed to flow beneath us.'

But Hal and his crew were not afraid. It had happened several times during their studies and on every occasion they found sperm whales to be gentle, curious animals. It was as if their vessel was included in the socialising.

An earlier account of sperm whale encounter shows a similar pattern, but it is always advisable to gain acceptance gradually, rather than pressing the whales for attention.

Eric Morris was cruising between two atolls in the Tuamotu Archipelago.

---

* During the Indian Ocean study the team swam with blue whales and filmed them. In the Gulf of St Lawrence the first underwater footage of blue whales, the largest creatures on earth (30 metres), was shot in August 1979.

The wind was so light his 12-metre Searunner trimaran was ghosting along at less than one knot. He sighted the white plumes of the whales half a kilometre to leeward. Half an hour later he saw them alter course. Moving lazily over the surface, curving towards him, the pod was led by two large male bulls. From the bulbous head and distinctive spouts he knew these were sperm whales. One of the bulls did a deep dive towards him while a group of cows and calves began to frolic just ahead. Eric lowered his sails and to his own amazement found himself preparing to snorkel with the whales. On entering the water, he found a huge tail hovering directly beneath him. It was the bull that had disappeared in a deep dive minutes before. It was floating motionless, head down, body vertical. When it moved its tail and swung its head towards Eric, he leapt aboard his boat in fear. As the rest of the whales slowly approached, his doubts quietened and he got back in. One of the cows swam close. Eric again sought refuge on his boat. The whale then stood vertically in the water lifting her head clear to look at him, revealing her long, thin lower jaw.

When the whales left, Eric motored after them, keen for more contact. This time he snorkelled to within 16 metres of a female with three calves around her. They appeared to bolt in fear. A bull advanced. From his boat Eric watched it gnash its jaws and thrash its tail with loud cracks. It seemed clear that he had gone too far in pressing his company on them and he left. (*Multihulls* magazine, November 1979.)

South Pacific Whale Encounters

In the South Pacific there have been a number of brief whale encounters with divers. A typical incident was when Ruth Hill and two companions approached a humpback mother and calf at Niue in August 1986. 'The mother was obviously wary of us but intensely curious.'

At 20 metres range the whale circled the divers and then left. On a second approach both divers and whales were more relaxed and they got within five metres of the mother as she eyed them, motionless, for several minutes. Then the calf descended to its mother, touched 'noses' and they left.

In the case of John Stoneman, veteran cameraman, a humpback approached him while alone. 'The animal swam up quite calmly and slowly, turned its massive head when three metres away, and looked at me through a large, brown eye.' When other divers approached, the whale accelerated away.

Peter Yates was on the nuclear protest ship *Fri* near Tahiti in May 1983. When a Bryde's whale, *Balaenoptera edeni*, approached the ship, six crew swam out to it. As the whale came close to Peter he felt a sudden rush of fear, but at that moment the whale dived deep and then surfaced to rise out of the water. Peter felt it communicated self-control and his groundless anxiety was settled.

Near the Galapagos Islands on 11 January 1978, a Bryde's whale surfaced alongside the 20-metre schooner of David and Gundi Day. Fearing a collision, David slowed to three knots but the whale began to frolic on their bow, performing a dozen barrel rolls. Then it ducked under the hull and took an interest in the dinghy trailing astern. Gundi got in the boat and the whale blew alongside. Twice she rubbed its lips and dark pieces of skin stuck to her hand. Then the whale lifted the dinghy gently on its nose, as if in response to her touch. It held its head even higher, as if attempting to look over the schooner's stern. After nearly two hours interaction, the whale appeared to tire and left. (*Marine Mammal News.*)

Close Approaches in Australia

Along Australia's Great Barrier Reef minke whales, *Balaenoptera acutorostrata*, have been mingling with skindivers for unusually long sessions. Marine biologist, Bill Gladstone, had several encounters near the Lizard Island Research Station over three seasons (1980-82). The encounters with large, solitary whales (9m) were brief, but with schools of smaller minke (6-8m), he had good contact for up to 35 minutes. Minke feed in antarctic waters, where they are still being hunted, and come to the Barrier Reef to breed during winter months (May to July).

On 23 June 1983, Bill Gladstone floated on the surface as five minke circled just below him, ascending at intervals to breathe. As each whale rose, Bill swam alongside making eye contact and wondering why these creatures were so gentle and curious towards man when compared with terrestrial animals. On this occasion, the encounter was cut short when aggressive silver tip sharks charged in and began circling. The whales were not troubled and continued swimming around the boat, raising their heads to look at Bill for a further 40 minutes.

At Marion Reef in August 1985, Irvin Rockman, Lord Mayor of Melbourne, spent five hours snorkelling at close quarters with eight medium-sized minke (7.6m). At 22 metres in coral country, the minke cavorted like playful puppies making shrill squeaks. Irvin shot several rolls of film and had the best diving experience of his life.

At Little Kelso Reef off Townsville, Alan Johnson's charter boat lay at anchor in 12 metres of water while a six-metre minke played around it for four hours and snorkellers and scuba-divers swam within touching distance.

Warrnambool had not seen right whales for a hundred years. Whalers drove them from their favourite mating grounds in the sheltered coves and bays of southern Victoria. Such gentle, slow moving creatures were easy targets.

Then, in 1982, seven right whales appeared in the shallows off Logan Beach, including a cow and calf. Each year, from that time, the whales return to this nursery area (and at Albany in West Australia), and the public turn out in droves to watch them from a specially erected look-out platform.

Divers venture out through the breakers to meet the whales but soon find that encounter is entirely the whales' choice, not theirs. 'Wait, until they they come to you,' was the message Allison and Peter Barker gained.

Out beyond the surf, three divers waited on the surface in the path of a mother and calf. As they came closer, the calf kept raising its head as if to see the objects in its way. Within visual range, the whales circled the divers. Allison was wondering where they had gone: 'Then I looked down. The green water became black. It was the mother's gigantic head. She swam straight beneath me, just under the surface.'

As the whale passed, the two men touched it with a fin or a hand. The whales circled once more and left.

In June 1969, at Algoa Bay, South Africa, Anne Rennie and David Allen met a right whale in the surf zone. They stroked its sides; David climbed on its back and Anne was towed several metres holding its tail. Other divers report similar experiences in this area, another right whale nursery.

At Portland Harbour in Victoria a rare pygmy right whale, *Caperea marginata*, began circling off the wharf on 28 November 1986. A young female, Rosie, probably only recently weaned, proved to be in good health. When she left after five weeks she was in perfect condition.

During her stay Rosie demonstrated a gentle nature and curiosity towards people who approached. Such was her willingness to mimic the actions of divers that she soon became a significant tourist attraction. Two at a time, divers would enter, lie motionless on the water, and the whale would rise, nudge them, nodding its head and grunting. She was extensively studied and photographed and her sounds were recorded by Dr Bill Dawbin of the Australian Museum.

Whale Encounters in New Zealand

In New Zealand our first good record of a close approach came from Dr Tony Ayling at Leigh Marine Laboratory. On 2 February 1978, four biologists were operating from boats off Goat Island when a 12-metre sei whale, *Balaenoptera borealis*, surfaced nearby. Four times the diving scientists leapt in with it and the whale showed no disquiet. On one pass it stopped just a metre below them. On another it turned on its side and looked up. As the divers left the whale broached three times with awe-inspiring crashes.

The most sensational close approach in New Zealand history produced a news release that went world-wide. In November 1980, the crew of the 25-metre purse-seiner, *Pirimai*, were alarmed when a 15-metre southern right whale approached their vessel, but when it returned two weeks later, they decided to accept its friendly advances. It was a glassy calm day off the Kaikoura coast of the South Island. Within metres of the ship the whale poked its head out of the water and examined the wooden fishing-boat, almost within touching distance.

The skipper, Chris Sharp, donned basic diving gear and joined it in the water. Several metres down, a huge eye scrutinised him only centimetres away. The human and the whale stared at each other and then both surfaced to breathe. The crew watched in amazement as the massive form swam slowly past their skipper, carefully twisting its tail to avoid contact. Two more of the crew decided to join him. Deckhand, Simon Reid, was lifted half out of the water on the huge tail.

Peter Kelliher jumped in and swam over: 'The whale was arching through the water and lifted me clear of the sea before gliding on out from under me.' Almost every newspaper in New Zealand carried their picture on its front page: Peter astride the cetacean's back, right alongside the fishing vessel.

To the press, Chris Sharp remarked: 'There is no doubt the whale was communicating with us. It showed absolutely no aggression and made no sudden movements — even when we jumped in right next to it. How anybody could kill such a unique, intelligent animal is impossible to comprehend.'

Pilot Whales — On 14 February 1984, our catamaran RV *Interlock* met an aggregation of longfin pilot whales, *Globicephala melaena*, and bottlenose dolphins near Whangaroa in the far north. From their behaviour we think they were feeding cooperatively on squid. While they were close around our vessel we recorded their noisy whistling and click chains. Then we played synthesised click chains up and down scale, through our sound array, listening for any responses. To our utter incredulity, Jan and I heard and recorded perfect runs up the scale, exactly mimicking our output. We do not know whether this was whale or dolphin, but suspect the former. It does indicate that echolocation clicks can be adapted to playful or communicative ends.

As the whales left, they paraded past our vessel on the surface, a grand procession of subgroups. Finally the largest whale of all, a male some six metres long, came alongside the bow where Jan stood, made eye contact with her and then scrutinised our underwater speaker before leaving with the others. Our recordings substantiate what we heard.

Out from Falmouth, England, the James Wharram Dolphin Research team met seven longfin pilot whales on 31 August 1984. The whales showed interest in their 16-metre catamaran *Tehini*. As Hanneke Boon was towed slowly between the bows the whales came within a metre. At one time they all waited ahead for *Tehini* to catch up.

On 8 July 1987, 24 kilometres out in the Mediterranean from Toulon, our French friend Franck Charreire had a musical meeting with a dozen longfin pilot whales — males, females and juveniles. Crouched on the sloping stern of his slow-moving catamaran, feet in the water, Frank played the flute while whales came within touching distance and made eye contact. They became increasingly playful, making blowhole noises in response to the flute, blowing bubbles, rolling belly up, and waving their flippers like a baby in a cradle. Meanwhile the crew filmed and recorded action and sounds until darkness closed in. Even then, they could hear the whales blowing around them.

Gentle Jaws — The most extraordinary sequence of close encounters Project Interlock has documented so far involves orca mouthing the legs of New Zealand divers. These anecdotes belong in a certain benign context: there had been several publicised episodes of friendly interaction with boats, such as the female who rested her head on a stern platform to be stroked; a film-maker who obtained footage of three orca head-standing in the shallows; and in separate incidents, divers who entered the water to release orca enmeshed in nets. But what ensued is almost beyond belief — if it had not occurred on five separate occasions.

In late December 1983, Peter O'Donnell was completing a day's lobster diving. On the pinnacle reef which rises from 22 metres of water out in Helena Bay, Northland, it was getting too dark to see. Peter called to his friend dozing in the boat about 150 metres away. Just then his foot hit something solid. He knew it could not be rock:

'I hesitated to look. I knew it must be something terrible. Next I felt whatever it was take hold of my flipper. I peered down. A great black and white thing was hovering there, vertical in the water, stock still. No sound or bubbles. I couldn't believe my eyes. After about ten seconds it just opened its mouth, slipped back and vanished in the murk.

'Then it started swimming back to me from different angles, gliding by at touching distance. I even thought of giving it my catch. Next it began rearing up out of the water alongside me, arching over. My companion looked over and thought I'd grabbed a sunfish. Slowly it dawned on him. He wrenched up the anchor. By the time the boat got there I'd been with the whale for about eight minutes. I'd given up on life. I just knew it was going to eat me. I bounded into the boat and we sat there stupefied as it started coming across and beneath us, almost touching the boat. Our boat was five metres long and I guess the orca was about seven metres in length. Thinking it over afterwards, I had a strong feeling that something really alien was trying to communicate with me — this incredibly strong alien mind. And I'm fairly sure now, that it wouldn't have hurt me.'

On 18 November 1984, David Wilson was scuba-diving just south of Helena Bay when something bumped his back. Thinking it was a rock David turned and found himself eyeball to eyeball with an orca. Bracing himself against a rock, David and the whale observed each other for one to two minutes. David prodded it gently with a piece of wood to keep it at a distance. After leaving David the orca leaped almost completely out of the water only a metre from the dive boat and its two occupants, as though it wanted to look in at them. A much larger orca was leaping a half-mile away.

The whale made a second approach to David, returned at speed to the boat, submerged a few metres short of a collision, and then returned to the diver a third time before rejoining its companion.

On 8 April 1985, Jim Skenars and Thelma Wilson were scuba-diving off New Plymouth where visibility was seven metres:

'An orca around four metres in length circled us at a metre range, regarding us with a rather large eye, going away after each circuit, presumably to breathe, and returning for another look.

'It made four passes inspecting us both together, then each separately, either circling closely or approaching head-on to look with both eyes. When its nose was next to us, the tail tended to disappear in the distance. It gave the impression of being curious and was not at all threatening. Thelma was able to run her hand down its side during one pass. We did feel particularly vulnerable but neither of us was keen to move away from the only large rock in the area, and must admit that we considered the possibility of fitting into a crack with the lobster. When we surfaced we discovered that our boat-lady had not seen a thing. Just as well or the boat may have found its way up the beach at 50 kph.'

Todd Sylvester, a biologist at Leigh Marine Laboratory, had been scuba-diving off Goat Island on 27 April 1985. Just as he surfaced he saw a five-metre-long orca about ten metres away. To be on the safe side he dived for the bottom, seeking cover.

'The next thing I knew, when I looked over my right shoulder, was the orca about a metre behind me! At this stage I still wasn't particularly alarmed, thinking it was just a chance encounter, with me going one way and the orca in another. I swam faster, but to my horror, the orca followed me. Things got infinitely worse when the orca opened its huge mouth. Rows of white teeth greeted me from about a metre from my feet. The orca then sped up and put its mouth around my right foot, ankle and flipper, *but at no stage did the whale close its mouth and bite me.* I took the chance and pulled my foot out of the cavernous mouth. No sooner had I done this, than the orca swam forward and put its mouth around my left foot and the calf region of my leg. Again, at no stage did it bite me. This time as well as pulling my leg out (I was swimming on my back), I "kicked" the orca off with my other foot. I was pretty excited, but wasn't panicking. Part of me thought I was going to become whale fodder, the other part hoped that the orca was only playing with me.

'The orca put its mouth around my lower leg region and I "kicked it off" about three or four times in the next minute or so. After the fifth or sixth mouthing, I had had enough so stopped kicking the whale off and dived quickly to the bottom. As it turned out, I wouldn't see the orca again. Thank God.'

These accounts were published in Project Interlock's newsletter 31 and circulated among the 10,000 readers of *Dive* magazine. We hoped that the next diver to meet a friendly orca would be less frightened and perhaps respond in a more positive manner.

On 30 August 1986, Gary Longley and two companions sighted eight orca out from Tauranga. When a male/female pair acted inquisitively, Gary decided to enter the water.

'Suddenly, out of the blue, one glided below me, tilted a little and stared up. It circled twice, moving closer to hang suspended a metre away, looking at me. Then sinking slightly, it opened its huge jaws and took the end of my flipper in its mouth. I recalled the newsletter accounts and my fears were relieved. Twice it gently mouthed my fin but with no attempt to bite me. Then the pair swam off.'

Encouraged by this safe outcome, Gary returned to the boat for a thawed baitfish. 'Lifting my head above water I saw a big fin coming towards me. Silently the two orca cruised by, eyeing me closely. On the next pass I waved the fish at them. Slowly one approached and mouthed it gently. Its massive teeth were clearly exposed only centimetres from my slightly trembling hand. Perhaps the stale baitfish wasn't quite to its liking because it swam away with me still holding the fish. Then the performance was repeated. Some very close passes ensued. Everything went dark as the beast gracefully swam within an arm's length. There was no turbulence, as I would have expected. Yet it was enormous. After several circuits the orca pair kept on with their eastward journey, ending the ten-minute encounter.'

Conclusion

'All the neurological evidence is not in, regarding the whale brain and intelligence. However, enough is known to lead us to believe we are dealing with special creatures with remarkably developed brains. Major riddles of nature and relations between species may indeed be answered by study of these brains, and these opportunities may die with the whales if we do not act now. They could have taught us much if we had only listened. Their kinship with man at the level of neurological development holds us in awe and fascination. It is unthinkable for us to sit idly by and let such unique beings wantonly be destroyed by selfish and short-sighted men. This is a resource and kinship that belongs to us all. Our very training and deepest feelings make us respect these wondrous creatures. Would that the brains of men could lead them to live in harmony with nature instead of ruthlessly plundering the seas that nurtured us . . . We must continue to be haunted by such solitary beings, with amazingly complex brains, wending their way through the seas, wondering perhaps, what manner of men are hunting them down to destroy them forever.'

— Dr Peter Morgane, Worcester Foundation for Experimental Biology, Worcester, Massachusetts.

Epilogue

I love forms beyond my own and regret the borders between us.
Loren Eiseley

Solitary dolphin episodes continue to unfold. As this book is prepared for press, four new stories have entered our files, but space precludes presentation in more than summary form. Others, as yet unreported, may be in progress. In 1988, we learnt of solitary dolphins near Barcelona, Spain; at Kotor, Yugoslavia and at Harbor Island, Bahamas; of 'Herbie', who only lets children touch him. Sadly, both the Romeo and Billy episodes have ended. Dorad is widely known as 'Funghi, the Dingle Dolphin'. Jean-Louis continues to entertain divers. For Fanny, a special observation post and research base has been established.

Dorad — At lonely Dingle Harbour in southwest Ireland, a male bottlenose dolphin became resident in mid 1984. When divers Brian Holmes and Sheila Stokes heard of it, they set about establishing a relationship, commencing in September 1986.

Few Dints have involved such intensive and exclusive contact. In the first month, they spent 14 days with the dolphin, diving three times each day for sessions of one-and-a-half hours minimum. Their relationship developed swiftly, initially with Sheila. Before long the dolphin was lifting them from below, pushing them halfway across the bay, or resting its beak quietly on their shoulders for several minutes at a time. The dolphin, Dorad, would sneak up from behind and butt them on the head; leap right over them and occasionally land on top of them.

Although they always spoke to him, for the initial period, Dorad was mute. Then one day, he suddenly rose in the water before Sheila and started squeaking nonstop. Since that time, he usually produced a variety of sounds. Both snorkel and scuba were used and on one occasion Dorad hit Brian's tank valve and cut his mouth; but he resumed friendly contact and swam with them right into the beach.

During the 1987 summer, novel gamesplay continued to develop, with Dorad giving tail-fin tows, coming up with a good-sized conger eel and goading them to chase him for the trophy. Another time Dorad gave Brian a garfish, caught another, then retrieved the first. He frequently released bubble-gulps in an obvious bid to gain attention.

During play sessions he was prone to defaecate or vomit over people in what seemed, to Brian, a sign of affection. Tooth rakings on his body indicate he has been meeting other cetaceans.

That July, Horace Dobbs made a second visit to Dorad, this time with a film crew to document the introduction of three mentally depressed people.

'It was extraordinary,' he wrote us. 'The dolphin appeared to become mesmerised by the oldest of the group.'

Horace knows there are too many depressed people in the world to introduce to wild dolphins, but his 'Operation Sunflower' aimed to open them up through film images of such encounters. He hopes this may convey that essential cathartic quality which dolphins have, and will produce the effect that recorded music does with sound: we need neither orchestra nor actual dolphin to be deeply moved.

Fanny — In mid-July 1987, a young female bottlenose began to linger around a beacon moored in 15 metres of water near the little Mediterranean port of Carro, just 40 kilometres from Marseille. Water police were first to notice her as she rode the bow wave of any vessel within half a mile of 'her' buoy.

Our Marseille friend, Franck Charreire, was delighted at her arrival. He had been hearing reports of another lone dolphin near Naples, but could not afford the travel. He observed as Fanny began to play with swimmers and divers; a new Jean-Louis situation was developing, with behaviours quite distinct from male solitary dolphins.

Although well protected by the local gendarmerie, who regarded her as a special charge, Franck organised an Information Day for Fanny. There was a press conference at which she was presented in the context of all other solitary dolphins; and also there was an official 'baptism', voted from the public.

The audience included public figures, scientists and the media, so that the dolphin's needs and welfare could be well established along with the inevitable flood of publicity.

Billy, the horse-training dolphin — In October 1987, news got out that a two-metre dolphin was living in Adelaide's Port River, Australia. It had arrived with its mother when about a metre long. The young dolphin befriended 60-year-old 'Sandy' Sandford, a horse trainer, who would exercise his gallopers behind a dinghy each day. The young dolphin, Billy, was intrigued but his mother kept her distance. Subsequently she left the river but Billy stayed. Before long, the dolphin was swimming right alongside the horses, often brushing against them. Amazingly, the horses accepted their companion, its dorsal cleaving the water by their heads. The story of the 'horse-training' dolphin continues . . .

Romeo — In 1985, a young male bottlenose began playing with people on several beaches 50 kilometres north of Naples, Italy. Around that time two other dolphins, possibly his companions, died in the area, one from swallowing a plastic bag, the other from a bullet wound, for which a man went to prison.

His territory was a 15-kilometre stretch of coast. He would play on certain beaches for several days and then move on, in an erratic pattern. He had no special friendships on the beaches and would often rub his belly on boats or flick water at the fishermen, inviting them to play.

Because of his vigorous sexual responses to women, the name Romeo was given. His sexual display was directed at men, too — the more people fondled him, the more he was aroused. Meeting a woman wearing a tampon he became highly excited and made a sonar scrutiny of it. When young men were a little too rough with him, Romeo soon put an end to it with a whack of his tail, which he found was more effective *above* water.

Our German and Swedish friends, Carola Hepp and Lars Lofgren, visited Romeo in October 1987. At Baia Domizia, Carola swam with the dolphin for long periods. She had a small silver ball with a bell inside. As soon as she entered the water Romeo would come to her side and tow her away from the crowds. She found it most comfortable lying on her back, holding his dorsal fin. At times he became highly aroused and tried to penetrate the defences of her tiny bikini but he never bit her as dolphins do during their matings.

She would often hear the soft clicking of his sonar and a high-pitch whistling — 'whenever he felt very happy'. Standing in the shallows, she played a game dodging his beak to left or right as he circled her rapidly. '*He blew big bubbles from his blowhole under water* as he moved towards me or whenever I pleased him.'

On her third day with Romeo, Carola had frolicked with him for three hours when she noticed a mood change. The dolphin became silent, circling slowly with no penile erection. When she touched his body, he twitched slightly. His eyes were half closed. He slowed even more so she could swim at his side. With one hand on his flank she dolphin-kicked and circled with him. She realised he was sleeping. Gradually she entered a meditative state and her pulse slowed. She felt calm and utterly safe as she rose and descended with the dolphin. Holding her breath longer than ever before she forgot time and place, not caring where she was going. With each circle they got further out to sea. She examined the scars on his body — a serious one on his back from a propeller, another deep one along one third of his right side. There were no tooth rakings to suggest contact with other dolphins.

Deeply entranced, Carola swam on and on with the dolphin. Then, arousing herself a little after a descent, she looked around. She was a kilometre offshore. When she started to swim back against the current, Romeo circled her all the way, still somnolent. Eventually her feet touched the sand. She saw beaches coming up — chairs, umbrellas, houses and people; young men stormed into the water to touch the dolphin and the girl that had arrived from the sea on to their beach.

Romeo woke and started pushing his beak under her feet. He didn't want her to walk up the beach. The second time he did this, she gently nailed his beak to the sand. He wriggled his head and blew bubbles. The game was on and he became rough. She responded in kind. He did all he could to prevent her leaving the sea. She lingered a few moments, kissing and hugging him. When she finally reached her amazed friends, Lars and Gina, exhaustion hit her. It was near dusk. The beach crowds had thinned. She had been in the sea, without a wetsuit, for five hours. A dolphin journey!

'As we sat in deck chairs talking, we saw him leap in the distance, five times against the sunset, high above the water in farewell. Then I became aware of a new perspective: everything that happens in our society is a game that human 'children' play — all the wars and politics and industries and our rotten games with nature. But I saw there was something that humans can not destroy and I felt a deep silence. There was hope within my heart.'

Project Interlock and the Reader

Now you are invited to join this adventure. In presenting interlock accounts from people like yourselves, who have had direct experiences with cetaceans, I hope you may discover patterns, develop insights or make interpretations which could lead to greater understanding of ocean mind. For this reason, I have minimised my own interpretations, conscious all the time that it would be possible to write another book from this material — comparing, analysing and drawing conclusions. But I have a hunch it is too early for this. The interspecies era has only just begun; it is best that the material is first shared with the global village.

Readers wishing to contact us or receive Project Interlock newsletters, which will continue reportage of encounters with cetaceans, should send a stamped, addressed envelope to us at P.O. Box 20, Whangarei, New Zealand. Readers in other countries should include a dollar US currency, or the equivalent. Donations to assist our research are always welcome. Our subsistence lifestyle makes postage the greatest outlay!

Encounters between divers and the Bahamas dolphins can be seen on a 30-minute V.H.S. format video obtainable from Larry Vertefay, Friends of the Sea, Box 2190, Enfield, Conn. 06082, U.S.A.